THE WORLD BOOK ENCYCLOPEDIA OF
PEOPLE AND PLACES

3
G-J

WORLD
BOOK

a Scott Fetzer company
Chicago
www.worldbookonline.com

For information about other World Book publications, visit our website at http://www.worldbookonline.com or call 1-800-WORLDBK (1-800-967-5325).

For information about sales to schools and libraries, call 1-800-975-3250 (United States); 1-800-837-5365 (Canada).

Library of Congress Cataloging-in-Publication Data

The World Book encyclopedia of people and places.
    v. cm.
    Summary: "A 7-volume illustrated, alphabetically arranged set that presents profiles of individual nations and other political/geographical units, including an overview of history, geography, economy, people, culture, and government of each. Includes a history of the settlement of each world region based on archaeological findings; a cumulative index; and Web resources"--Provided by publisher.
    Includes index.
    ISBN 978-0-7166-3758-5
    1. Encyclopedias and dictionaries. 2. Geography--Encyclopedias. I. World Book, Inc. Title: Encyclopedia of people and places.
    AE5.W563 2011
    030--dc22
                                                    2010011919

This edition ISBN: 978-0-7166-3760-8

Printed in Hong Kong by Toppan Printing Co. (H.K.) LTD
3rd printing, revised, August 2012

Cover image:
Jain Temple, Jaisalmer, Rajasthan, India

© Margo Silver, Stone/ Getty Images

CONTENTS

POLITICAL WORLD MAP

The world has 196 independent countries and about 50 dependencies. An independent country controls its own affairs. Dependencies are controlled in some way by independent countries. In most cases, an independent country is responsible for the dependency's foreign relations and defense, and some of the dependency's local affairs. However, many dependencies have complete control of their local affairs.

By 2010, the world's population was nearly 7 billion. Almost all of the world's people live in independent countries. Only about 13 million people live in dependencies.

Some regions of the world, including Antarctica and certain desert areas, have no permanent population. The most densely populated regions of the world are in Europe and in southern and eastern Asia. The world's largest country in terms of population is China, which has more than 1.3 billion people. The independent country with the smallest population is Vatican City, with only about 830 people. Vatican City, covering only 1/6 square mile (0.4 square kilometer), is also the smallest in terms of size. The world's largest nation in terms of area is Russia, which covers 6,601,669 square miles (17,098,242 square kilometers).

Every nation depends on other nations in some way. The interdependence of the entire world and its peoples is called *globalism*. Nations trade with one another to earn money and to obtain manufactured goods or the natural resources that they lack. Nations with similar interests and political beliefs may pledge to support one another in case of war. Developed countries provide developing nations with financial aid and technical assistance. Such aid strengthens trade as well as defense ties.

4

Nations of the World

Name	Map key		Name	Map key		Name	Map key	
Afghanistan	D	13	Bulgaria	C	11	Dominican Republic	E	6
Albania	C	11	Burkina Faso	E	9	East Timor	F	16
Algeria	D	10	Burundi	F	11	Ecuador	F	6
Andorra	C	10‡	Cambodia	E	15	Egypt	D	11
Angola	F	10	Cameroon	E	10	El Salvador	E	5
Antigua and Barbuda	E	6	Canada	C	4	Equatorial Guinea	E	10
Argentina	G	6	Cape Verde	E	8	Eritrea	E	12
Armenia	D	12	Central African Republic	E	10	Estonia	C	11
Australia	G	16	Chad	E	10	Ethiopia	E	11
Austria	C	10	Chile	G	6	Federated States of Micronesia	E	17
Azerbaijan	D	12	China	D	14	Fiji	F	1
Bahamas	D	6	Colombia	E	6	Finland	B	11
Bahrain	D	12	Comoros	F	12	France	C	10
Bangladesh	D	14	Congo, Democratic Republic of the	F	11	Gabon	F	10
Barbados	E	7	Congo, Republic of the	F	10	Gambia	E	9
Belarus	C	11	Costa Rica	E	5	Georgia	C	12
Belgium	C	10	Côte d'Ivoire	E	9	Germany	C	10
Belize	E	5	Croatia	C	10	Ghana	E	9
Benin	E	10	Cuba	D	5	Greece	D	11
Bhutan	D	14	Cyprus	D	11	Grenada	E	6
Bolivia	F	6	Czech Republic	C	10	Guatemala	E	5
Bosnia-Herzegovina	C	10	Denmark	C	10	Guinea	E	9
Botswana	G	11	Djibouti	E	12	Guinea-Bissau	E	9
Brazil	F	7	Dominica	E	6			
Brunei	E	15						

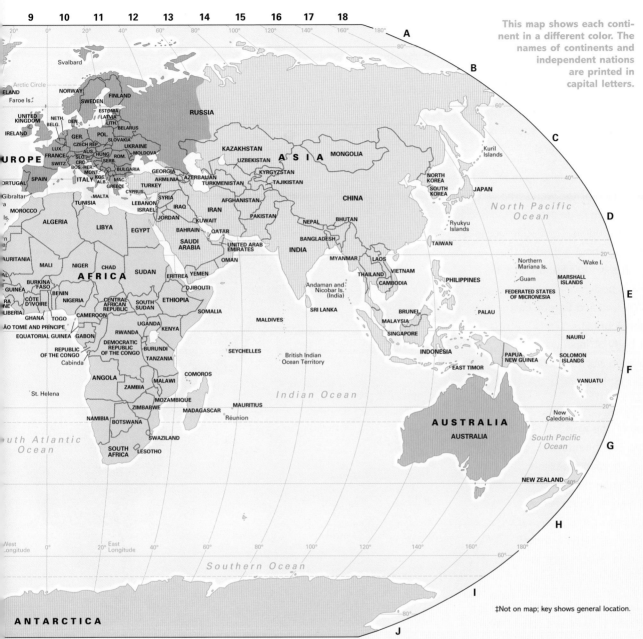

This map shows each continent in a different color. The names of continents and independent nations are printed in capital letters.

‡Not on map; key shows general location.

Name	Map key		Name	Map key		Name	Map key		Name	Map key		Name	Map key	
Guyana	E	7	Lebanon	D	11	Namibia	G	10	St. Vincent and the Grenadines	E	6	Taiwan	D	16
Haiti	E	6	Lesotho	G	11	Nauru	F	18	Samoa	F	1	Tajikistan	D	14
Honduras	E	5	Liberia	E	9	Nepal	D	14	San Marino	C	10‡	Tanzania	F	11
Hungary	C	10	Libya	D	10	Netherlands	C	10	São Tomé and Príncipe	E	10	Thailand	E	15
Iceland	B	9	Liechtenstein	C	10‡	New Zealand	G	18	Saudi Arabia	D	12	Togo	E	9
India	D	13	Lithuania	C	11	Nicaragua	E	5	Senegal	E	9	Tonga	F	1
Indonesia	F	16	Luxembourg	C	10	Niger	E	10	Serbia	C	10	Trinidad and Tobago	E	6
Iran	D	12	Macedonia	C	11	Nigeria	E	10	Seychelles	F	12	Tunisia	D	10
Iraq	D	12	Madagascar	F	12	Norway	B	10	Sierra Leone	E	9	Turkey	D	11
Ireland	C	9	Malawi	F	11	Oman	E	12	Singapore	E	15	Turkmenistan	D	13
Israel	D	11	Malaysia	E	15	Pakistan	D	13	Slovakia	C	11	Tuvalu	F	1
Italy	C	10	Maldives	E	13	Palau	E	16	Slovenia	C	11	Uganda	E	11
Jamaica	E	6	Mali	E	9	Panama	E	5	Solomon Islands	F	18	Ukraine	C	11
Japan	D	16	Malta	D	10	Papua New Guinea	F	17	Somalia	E	12	United Arab Emirates	D	12
Jordan	D	11	Marshall Islands	E	18	Paraguay	G	7	South Africa	G	11	United Kingdom	C	9
Kazakhstan	C	13	Mauritania	D	9	Peru	F	6	Spain	C	9	United States	C	4
Kenya	E	11	Mauritius	F	12	Philippines	E	16	Sri Lanka	E	14	Uruguay	G	7
Kiribati	F	1	Mexico	D	4	Poland	C	10	Sudan	E	11	Uzbekistan	D	14
Korea, North	C	16	Moldova	C	11	Portugal	D	9	Sudan, South	E	11	Vanuatu	F	18
Korea, South	D	16	Monaco	C	10‡	Qatar	D	12	Suriname	E	7	Vatican City	C	10‡
Kosovo	C	11	Mongolia	C	13	Romania	C	11	Swaziland	G	11	Venezuela	E	6
Kuwait	D	12	Montenegro	C	10	Russia	C	13	Sweden	B	10	Vietnam	E	15
Kyrgyzstan	C	13	Morocco	D	9	Rwanda	F	11	Switzerland	C	10	Yemen	E	12
Laos	E	15	Mozambique	F	11	St. Kitts and Nevis	E	6	Syria	D	11	Zambia	F	11
Latvia	C	11	Myanmar	D	14	St. Lucia	E	6				Zimbabwe	G	11

The surface area of the world totals about 196,900,000 square miles (510,000,000 square kilometers). Water covers about 139,700,000 square miles (362,000,000 square kilometers), or 71 percent of the world's surface. Only 29 percent of the world's surface consists of land, which covers about 57,200,000 square miles (148,000,000 square kilometers).

Oceans, lakes, and rivers make up most of the water that covers the surface of the world. The water surface consists chiefly of three large oceans—the Pacific, the Atlantic, and the Indian. The Pacific Ocean is the largest, covering about a third of the world's surface. The world's largest lake is the Caspian Sea, a body of salt water that lies between Asia and Europe east of the Caucasus Mountains. The world's largest body of fresh water is the Great Lakes in North America. The longest river in the world is the Nile in Africa.

The land area of the world consists of seven continents and many thousands of islands. Asia is the largest continent, followed by Africa, North America, South America, Antarctica, Europe, and Australia. Geographers sometimes refer to Europe and Asia as one continent called Eurasia.

The world's land surface includes mountains, plateaus, hills, valleys, and plains. Relatively few people live in mountainous areas or on high plateaus since they are generally too cold, rugged, or dry for comfortable living or for crop farming. The majority of the world's people live on plains or in hilly regions. Most plains and hilly regions have excellent soil and an abundant water supply. They are good regions for farming, manufacturing, and trade. Many areas unsuitable for farming have other valuable resources. Mountainous regions, for example, have plentiful minerals, and some desert areas, especially in the Middle East, have large deposits of petroleum.

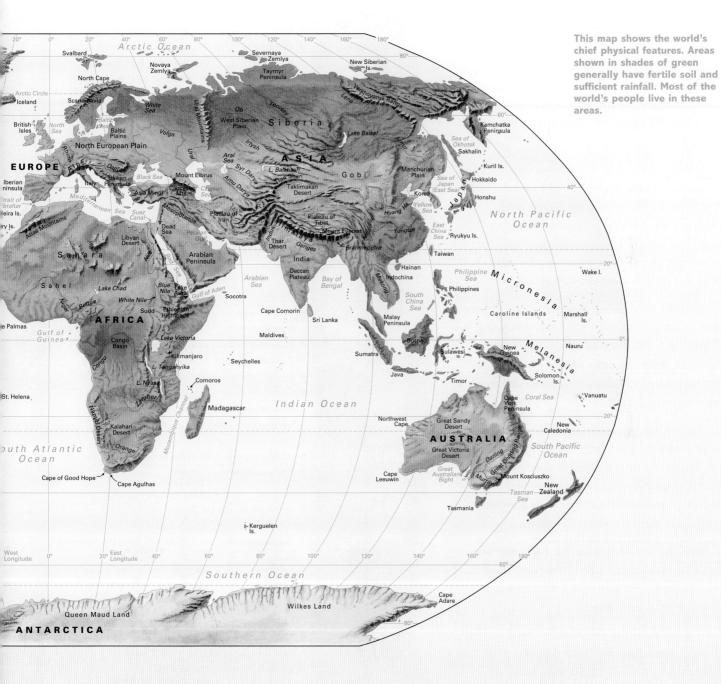

This map shows the world's chief physical features. Areas shown in shades of green generally have fertile soil and sufficient rainfall. Most of the world's people live in these areas.

GABON

The small, forested country of Gabon is one of the most thinly populated nations in Africa. It has a population of about 1,394,000 people. Most of Gabon's people are farmers who live in small villages on the Atlantic coast, along the Ogooué River or one of its branches, or in the less forested north. One of the inland river towns, called Lambaréné, became known worldwide as the home of Albert Schweitzer—an acclaimed Alsatian doctor and missionary who built a hospital and leper colony near there in the early 1900's.

Most of the people of Gabon once lived in houses with walls made of mud-covered branches and thatched roofs. Today, many houses have corrugated metal roofs and cement walls. Each Gabonese village usually has a meeting place where the older men gather to visit and discuss community affairs.

The people of Gabon represent many African ethnic groups. The most important group, the Fang, live in the north and were once fierce warriors, feared by other African ethnic groups as well as by Europeans. Today, the Fang dominate national politics.

A small but important group of related peoples are the Omyéné, who live along the coast of Gabon and were the first to meet and deal with European traders and missionaries. These activities gave the Omyéné an early advantage in trade and education in the country.

Small groups of Pygmies hunt and trap animals for food in the southern forests of Gabon. The Pygmies were one of the first peoples to live in what is now Gabon.

The first Europeans to reach the region were Portuguese sailors who landed on the coast sometime in the 1470's. Europeans carried on a slave trade with the Omyéné people who lived along the coast, and in 1839 France established a naval and trading post there.

FACTS

Official name:	Republique Gabonaise (Gabonese Republic)
Capital:	Libreville
Terrain:	Narrow coastal plain; hilly interior; savanna in east and south
Area:	103,347 mi² (267,667 km²)
Climate:	Tropical; always hot, humid
Main river:	Ogooué and its branches
Highest elevation:	Mont Iboundji, 5,167 ft (1,575 m)
Lowest elevation:	Atlantic Ocean, sea level
Form of government:	Republic
Head of state:	President
Head of government:	Prime minister
Administrative areas:	9 provinces
Legislature:	Senate with 102 members serving six-year terms and the Assemblée Nationale (National Assembly) with 120 members serving five-year terms
Court system:	Cour Supreme (Supreme Court), Constitutional Court, Courts of Appeal, lower courts
Armed forces:	4,700 troops
National holiday:	Independence Day - August 17 (1960)
Estimated 2010 population:	1,394,000
Population density:	13 persons per mi² (5 per km²)
Population distribution:	84% urban, 16% rural
Life expectancy in years:	Male, 54; female, 56
Doctors per 1,000 people:	0.3
Birth rate per 1,000:	27
Death rate per 1,000:	12
Infant mortality:	58 deaths per 1,000 live births
Age structure:	0-14: 36%; 15-64: 69%; 65 and over: 5%
Internet users per 100 people:	8
Internet code:	.ga
Languages spoken:	French (official), Fang, Myene, Nzebi, Bapounou/Eschira, Bandjabi
Religions:	Christian, Muslim, animist
Currency:	Coopération Financière en Afrique Centrale franc
Gross domestic product (GDP) in 2008:	$14.96 billion U.S.
Real annual growth rate (2008):	3.6%
GDP per capita (2008):	$10,264 U.S.
Goods exported:	Mostly: crude oil Also: manganese, timber
Goods imported:	Chemicals, food, machinery, transportation equipment
Trading partners:	China, France, United States

The Mont-Bouet Market is the largest market in Libreville, Gabon's capital, largest city, and economic center.

Gabon is a western African country that straddles the equator, just south of the continent's bulge into the Atlantic Ocean. Most of Gabon lies in the Ogooué River Basin.

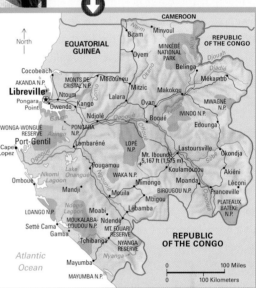

Soon, missionaries arrived and opened schools. In 1849, a group of former slaves landed at the trading post, which was then named *Libreville* (Free Town) and is now the capital of Gabon. The French also explored the inland rivers and developed a trade in lumber. In 1910, Gabon became part of French Equatorial Africa, and French companies acquired a great deal of land—as well as control over Gabon's foreign trade and forest products.

After World War II (1939–1945), Gabon began to move toward independence, and on Aug. 17, 1960, it became a free nation. Leon Mba, who had headed the government since 1957, became president.

In January 1964, Mba dissolved the national legislature. He was arrested by army officers a month later in an attempt to overthrow his government. But French troops came to Mba's aid, crushed the revolt, and restored Mba to power. Mba was reelected in 1967, but he died that same year. He was succeeded by Vice President Bernard-Albert Bongo, who later changed his name to El Hadj Omar Bongo.

Bongo was elected in 1973 and again in 1979 and 1986. In 1993, multiparty elections were held for the first time. Bongo was still elected and was reelected in 1998 and 2005. After Bongo's death in 2009, voters elected his son, Ali Ben Bongo, to succeed him.

Today, the president is still the country's most powerful official. A prime minister carries out the daily operations of the government.

LAND AND ECONOMY

Gabon lies directly on the equator, and much of its land is covered with dense forests. Thousands of plant species, including hundreds of species of trees, thrive in the hot, humid tropical climate.

The average annual temperature in Gabon is about 79° F (26° C). Rainfall is heavy throughout Gabon, especially along its northern coast on the Atlantic. There, the cold Benguela Current from the south meets the warm Guinea Current from the north, condensing moisture in the air into rain.

Much of the coast is lined with palm-fringed beaches, swamps, and lagoons, and large areas are covered with tall papyrus grass. Inland, the land gradually rises to forested rolling hills and low mountain ranges. Most of Gabon lies in the basin of the Ogooué River, which has cut many valleys through these highlands.

Agriculture and forestry

The majority of the Gabonese people live along the coast or the rivers and make their living by farming. They clear the forest around their villages and plant bananas, cassava, and yams—their main food crops. They also grow mangoes, oranges, and pineapples. Some farmers raise livestock, and many catch fish in the rivers or hunt animals in the forests. On the fertile land of northwestern Gabon, many farmers grow cacao and coffee.

Heavily forested Gabon is one of the richest countries in Africa in terms of resources. Its forests have long been its chief source of wealth, providing high-quality lumber for more than 100 years.

The lumber trade was developed in the mid-1800's when the French paddled up the Ogooué River into the interior. French companies eventually acquired control of the Gabonese forestry industry, and even now, much of Gabon's economy is controlled by French companies.

Today, lumber is one of the country's main exports. Wood from the huge okoumé trees that grow in Gabon is used to make plywood, and the lush forests also produce ebony and mahogany, valuable dark hardwoods.

A Gabonese man loads cassava onto a small boat at Port-Gentil. Cassava is one of the country's major food crops, but very little is grown for export.

The Trans-Gabon Railway was built to help the development of the mining industry in Gabon. It connects Franceville in the interior with the port of Owendo.

Port-Gentil, a major Gabonese port, lies on the Gulf of Guinea, an arm of the Atlantic Ocean, near the mouth of the Ogooué River. The river is still one of the chief means of transportation in the country.

Mining

Gabon has rich deposits of iron and manganese, and mining has become increasingly important to the Gabonese economy. In addition to iron and manganese, uranium and petroleum are also exported. Petroleum is Gabon's leading export. Gabon is a member of the Organization of the Petroleum Exporting Countries (OPEC).

An improving standard of living

In the past, most of the Gabonese people were poor, and their standard of living was correspondingly low. Since the mid-1900's, however, the development of their country's mineral resources has helped raise living standards. Mining provides jobs, and the income from mining has helped pay for social services throughout the country.

Unlike many African nations, the great majority of children in Gabon go to primary school, and the number attending secondary school is increasing rapidly. Since the early 1950's, the number of schools run by the government as well as by churches has also increased considerably. Today, over half of the adults can read and write.

Felled by a forester, one of the huge tropical trees that grow in Gabon crashes to the ground. Lumber is one of the country's main exports. Gabon's forests have provided high-quality lumber for more than 100 years.

Thick forests hug the beach in many coastal areas of Gabon. Such hardwoods as ebony and mahogany thrive in the hot, rainy equatorial climate.

Roads built during the mid-1900's link various parts of Gabon, but one of the chief means of transportation is still the Ogooué River. In the 1970's and 1980's, the Trans-Gabon Railroad was built, linking the port of Owendo with the interior. Improved railroads can help Gabon mine rich mineral deposits in remote areas. Some of the outlying regions in the north, east, and southwest must export their products through the neighboring countries of Cameroon and Congo.

Libreville, the capital city, is also a major port and the center of the country's commerce and culture. Since the late 1900's, a convention center and many other buildings have been erected in the city. And, although manufacturing in Gabon is limited, factory workers in Libreville produce food products, furniture, lumber, and textiles.

GAMBIA

Gambia is one of the smallest nations in Africa. The country lies on a narrow strip of land only 15 to 30 miles (24 to 48 kilometers) wide and 180 miles (290 kilometers) long on either side of the Gambia River. Gambia has a population of about 1,847,000 and an area of 4,361 square miles (11,295 square kilometers). Except for its short Atlantic coast, Gambia is entirely surrounded by Senegal.

Government and economy

Gambia is a republic and a member of the Commonwealth of Nations. The people elect a president to a five-year term. They also elect the majority of members of the National Assembly. The president appoints members of the Cabinet.

Gambia is a poor country with no valuable mineral deposits and little fertile soil. Mangrove swamps line the banks of the Gambia River from the coast to the center of the country. Beyond these swamps lie the *banto faros,* areas of ground that are firm during the dry season but turn into swamps when it rains. Near the ocean, the river water mixes with salt water and floods the banto faros, ruining the soil for crops. But farther inland, areas that the river floods with fresh water are used to raise rice.

Beyond the banto faros lie sandy plateaus, where farmers raise rice and peanuts. Rice, millet, and corn are grown mainly by Gambians as food for their families. Most Gambian farmers earn their living by raising peanuts, which have been exported from Gambia since 1830 and continue to be an important export crop. Construction and tourism also make important contributions to the economy.

People

Almost all Gambians are Africans and Muslims. They belong mainly to five ethnic groups: the Mandinka, the Fula (or Fulani), the Wolof, the Serahuli, and the Jola.

The Mandinka live throughout Gambia and work as traders and peanut farmers. Most of the Fula live in eastern Gambia and raise cattle and grow crops for a living. Many of the Fula are nomadic people. The Wolof are mainly northern farmers or residents of Banjul, the capital. Most of the Serahuli are farmers in the east, where the soil

FACTS

Official name:	**Republic of the Gambia**
Capital:	**Banjul**
Terrain:	**Flood plain of the Gambia River flanked by some low hills**
Area:	**4,361 mi² (11,295 km²)**
Climate:	**Tropical; hot, rainy season (June to November); cooler, dry season (November to May)**
Main river:	**Gambia**
Highest elevation:	**Unnamed location, 164 ft (50 m)**
Lowest elevation:	**Atlantic Ocean, sea level**
Form of government:	**Republic**
Head of state:	**President**
Head of government:	**President**
Administrative areas:	**5 divisions, 1 city**
Legislature:	**National Assembly with 53 members serving five-year terms**
Court system:	**Supreme Court**
Armed forces:	**800 troops**
National holiday:	**Independence Day - February 18 (1965)**
Estimated 2010 population:	**1,847,000**
Population density:	**424 persons per mi² (164 per km²)**
Population distribution:	**55% urban, 45% rural**
Life expectancy in years:	**Male, 55; female, 58**
Doctors per 1,000 people:	**0.1**
Birth rate per 1,000:	**38**
Death rate per 1,000:	**11**
Infant mortality:	**82 deaths per 1,000 live births**
Age structure:	**0-14: 42%; 15-64: 55%; 65 and over: 3%**
Internet users per 100 people:	**7**
Internet code:	**.gm**
Languages spoken:	**English (official), Mandinka, Wolof, Fula, other indigenous vernaculars**
Religions:	**Muslim 90%, Christian 8%, indigenous beliefs 2%**
Currency:	**Dalasi**
Gross domestic product (GDP) in 2008:	**$790 million U.S.**
Real annual growth rate (2008):	**5.5%**
GDP per capita (2008):	**$499 U.S.**
Goods exported:	**Fish, fruits and vegetables, peanut products**
Goods imported:	**Food, motor vehicles, petroleum products**
Trading partners:	**China, Denmark, Germany, Senegal, United Kingdom**

The Albert Market is a popular open-air street market in Banjul, the capital and largest city in Gambia. The market was built during the mid-1800's.

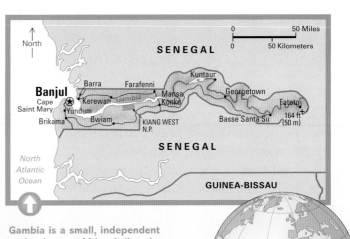

Gambia is a small, independent nation in west Africa. It lies along the Gambia River. Except for its short Atlantic coastline, Gambia is surrounded entirely by Senegal. Most of Gambia's land is flat and sandy, with mangrove and scrub forests along the coast and river.

is poor and agriculture is difficult. The Jola, once bitter enemies of the Mandinka, have lived in Gambia longer than the other ethnic groups. The Jola are southeastern farmers who raise rice and millet for food.

History

The area that is now Gambia was once part of the powerful west African Mali Empire. The English and Portuguese set up a slave trade in the area during the 1500's. All or part of the region was controlled by the English from the 1660's until 1965, when Gambia became independent.

On April 24, 1970, Gambia became a republic. David (now Sir Dawda, or Dauda) Jawara was elected president and reelected until a group of military officers overthrew him in 1994. They began ruling the country through a military council.

In 1996, Gambia adopted a new constitution, and Yahya Jammeh won presidential elections later that year. Soon after the elections, Jammeh dissolved the military council. In 1997, Gambia elected a National Assembly. Jammeh was reelected in 2001, 2006, and 2011. The Jammeh government made significant improvements in education, health, and agriculture.

Passengers board a steamer that carries people and goods on the Gambia River. Small oceangoing ships can sail up the Gambia as far as Kuntaur.

GEORGIA

Georgia, one of the former Soviet republics, is a mountainous nation on the eastern shore of the Black Sea. It has an area of 26,911 square miles (69,700 square kilometers). It is bordered by Turkey and Armenia in the south, Azerbaijan in the east, and Russia in the north. The population is about 4,292,000.

History

Wars and invasions mark much of Georgia's long history. The first Christian state was established in Georgia in the A.D. 300's. In the 1200's, the country was conquered by the Mongols, and in the 1300's by the fierce Tatar warrior called Tamerlane. From the 1500's to the 1700's, the Ottoman Empire and Iran struggled for control of the territory.

Between 1801 and 1829, Russia acquired all of Georgia. In 1918, after the October Revolution, the Georgian Menshevik Party proclaimed Georgia's independence. Although the government in Moscow recognized the declaration in 1920, the Red Army invaded Georgia in 1921, and it became a Soviet republic. In 1922, Georgia and the Soviet republics of Armenia and Azerbaijan formed the Transcaucasian Federation. The federation was one of the four original republics that joined to form the Soviet Union in 1922. In 1936, Georgia became a separate Soviet republic.

In the late 1980's, a strong independence movement emerged in Georgia. In April 1991, Georgia's Parliament declared Georgia independent. Later that year, the Soviet Union was dissolved.

Land and people

Georgia's varied landscape ranges from snow-capped mountains and dense forests to fertile valleys and warm, humid coastlands along the Black Sea. The Great Caucasus Mountains extend across northern Georgia, and the Little Caucasus Mountains cover the south. Farming is a major economic activity. Farmers raise beef and dairy cattle, hogs, poultry, and sheep. They also grow citrus fruit, corn, grapes, tea, and wheat.

FACTS

Official name:	Sak'art'velo (Georgia)
Capital:	Tbilisi
Terrain:	Largely mountainous; lowlands open to the Black Sea in the west; good soils in river valley flood plains
Area:	26,911 mi² (69,700 km²)
Climate:	Warm and pleasant; Mediterranean-like on Black Sea coast
Main rivers:	Kur, Rioni
Highest elevation:	Mount Shkhara, 17,163 ft (5,201 m)
Lowest elevation:	Black Sea, sea level
Form of government:	Republic
Head of state:	President
Head of government:	Prime minister
Administrative areas:	9 regions, 1 city, 2 autonomous republics
Legislature:	Umaghiesi Sabcho (Supreme Council or Parliament) with 150 members serving five-year terms
Court system:	Supreme Court, Constitutional Court
Armed forces:	21,200 troops
National holiday:	Independence Day - May 26 (1918)
Estimated 2010 population:	4,292,000
Population density:	159 persons per mi² (62 per km²)
Population distribution:	53% urban, 47% rural
Life expectancy in years:	Male, 72; female, 80
Doctors per 1,000 people:	4.7
Birth rate per 1,000:	11
Death rate per 1,000:	10
Infant mortality:	16 deaths per 1,000 live births
Age structure:	0-14: 18%; 15-64: 67%; 65 and over: 15%
Internet users per 100 people:	8
Internet code:	.ge
Languages spoken:	Georgian (official), Russian, Armenian, Azeri
Religions:	Orthodox Christian 83.9%, Muslim 9.9%, Armenian-Gregorian 3.9%, other 2.3%
Currency:	Lari
Gross domestic product (GDP) in 2008:	$12.98 billion U.S.
Real annual growth rate (2008):	2.4%
GDP per capita (2008):	$2,936 U.S.
Goods exported:	Copper, ferro-alloys, fruits and nuts, motor vehicles, scrap metals, wine
Goods imported:	Automobiles, machinery, petroleum products, pharmaceuticals, sugar, wheat
Trading partners:	Azerbaijan, Germany, Russia, Turkey, Ukraine, United States

Georgia manufactures chemicals, food products, and steel. Mines yield *barite* (barium ore), coal, copper, and manganese.

About 85 percent of Georgia's people are ethnic Georgians. The rest of the people are mainly Armenian, Abkhazian, Azerbaijani, Ossetian, and Russian.

Political problems

In 1991, Georgia elected Zviad K. Gamsakhurdia president. Opposition leaders, however, soon accused Gamsakhurdia of moving toward dictatorship. In 1992, opposition leaders formed an alternate government, and Gamsakhurdia fled the country. Eduard A. Shevardnadze became head of the State Council that ruled Georgia and later, chairman of Parliament.

Beginning in the early 1990's, tensions between Georgians and other ethnic groups turned violent. Conflict erupted in the regions of South Ossetia, in north-central Georgia, and Abkhazia, in northwestern Georgia. Ossetians and Georgians agreed to a cease-fire in 1992. The Abkhazians drove Georgian forces from Abkhazia by late 1993, but fighting erupted again in the early 2000's.

In 1995, a new constitution went into effect, and voters elected Shevardnadze president. In 2003, an uprising over disputed elections forced him to resign. The uprising became known as the "Rose Revolution" because the protesters carried roses as a symbol of nonviolence. Mikheil Saakashvili, a key leader in the uprising, was elected president in 2004 and reelected in 2008.

In August 2008, Georgia and Russia fought a five-day war over control of South Ossetia. Russian troops entered Georgia, and South Ossetia and other parts of Georgia suffered much damage. After the conflict, Russia recognized South Ossetia and Abkhazia as independent. The United States and other Western countries, however, continued to consider the areas as part of Georgia. Abkhazia and South Ossetia held presidential elections in 2011, which Georgia and other countries denounced as illegitimate.

A sea of poppies carpets the plains of Georgia. Georgia's greatest natural resources are its fertile soil and mild climate, which make farming a major economic activity.

Georgia lies mostly in southwestern Asia, but part of northern Georgia is located in Europe.

GERMANY

Germany's geographical location in central Europe has drawn many different peoples throughout history. The nation has no natural barriers to the east or west, and migrating groups often crossed the seas to the north. As a result, Germany has long served as a meeting place for the exchange of social, economic, and intellectual ideas. The country has also witnessed a long history of political conflicts.

Germany entered a period of political and economic crises following its defeat in World War I (1914-1918). In 1933, Adolf Hitler, leader of the Nazi Party, established a brutal dictatorship. Millions of Jews and others were murdered under Hitler's rule. Hitler led Germany into World War II (1939-1945), in which Germany was defeated.

From 1949 to 1990, Germany was divided into two separate countries: the Federal Republic of Germany (West Germany) and the German Democratic Republic (East Germany). West Germany became a parliamentary democracy with strong ties to Western Europe and the United States. East Germany became a Communist dictatorship closely associated with the Soviet Union.

In 1989, reform movements calling for increased political and economic freedom swept through the Communist nations of Europe. In response to these protests, the Berlin Wall, a symbol of the East German government's control over its citizens, was opened in November 1989. Free parliamentary elections were held in March 1990, and non-Communists gained control of the East German government.

On Oct. 3, 1990, East and West Germany were unified into a single nation. Berlin was made the capital of the united Federal Republic of Germany. Today, all Germans share the same freedoms, including freedom of speech, freedom of the press, and freedom to travel throughout a single Germany undivided by borders and guards.

The German people have a reputation for being hard-working and disciplined, but they are also known for their love of music, dancing, good food, and good fellowship. People from all over the world come to Germany to participate in the country's lively festivals. Many Germans also enjoy visiting the nation's scenic mountain areas and such popular tourist attractions as Neuschwanstein Castle.

The German people have made many important contributions to culture. Germany has produced such famous composers as Johann Sebastian Bach and Ludwig van Beethoven, and such literary giants as Johann Wolfgang von Goethe and Thomas Mann. In addition, German scientists have made significant breakthroughs in chemistry, medicine, and physics.

17

GERMANY TODAY

The merging of the economic, legal, and social systems of East and West Germany was officially completed on Oct. 3, 1990. Unity Day was welcomed with a mixture of joy and apprehension by the German people. Much of the population celebrated, but others believed the occasion called for more solemn contemplation. Many Germans were concerned about the costs of modernizing East Germany's crumbling economy.

International reaction was also mixed. While many world leaders declared their hope for a more united Europe, others expressed anxiety over the unification. Some feared the future military strength of a united Germany. Aware of such concerns, German leaders from both the East and West pledged that Germany would serve the world in a "peace-loving, freedom-loving" role.

Germany is a federal republic. The government's main bodies and offices include a Parliament, a federal chancellor, and a Cabinet.

Political unity

On Dec. 2, 1990, the first national elections after unification brought an easy victory to Helmut Kohl, who had led the drive for unity. Kohl, who was chancellor of West Germany from 1982 to 1990, headed a coalition of the Christian Democrats and Free Democrats. The 1990 elections were the first free elections held in a united Germany since 1932, when the Nazi Party was rising to power.

Following elections in 1998, Gerhard Schröder of the Social Democratic Party became Germany's chancellor. Angela Merkel of the Christian Democratic Union became chancellor in 2005. She became the first woman to hold the post.

People

Germany, with more than 82 million people, ranks second in population among the countries of Europe. Only Russia has more people. Berlin, the country's capital, is also its largest city. Most of the people of Germany live in urban areas.

FACTS

Official name:	Bundesrepublik Deutschland (Federal Republic of Germany)
Capital:	Berlin
Terrain:	Lowlands in north, uplands in center, Bavarian Alps in south
Area:	137,847 mi^2 (357,022 km^2)
Climate:	Temperate and marine; cool, cloudy, wet winters and summers; occasional warm foehn wind
Main rivers:	Rhine, Danube, Elbe, Weser, Oder
Highest elevation:	Zugspitze, 9,721 ft (2,963 m)
Lowest elevation:	Sea level along the coast
Form of government:	Federal Republic
Head of state:	President
Head of government:	Chancellor
Administrative areas:	16 Laender (states)
Legislature:	Parlament (Parliament) consisting of the Bundestag (Federal Assembly) with 614 members serving four-year terms and the Bundesrat (Federal Council) with up to 69 members
Court system:	Bundesverfassungsgericht (Federal Constitutional Court)
Armed forces:	244,300 troops
National holiday:	Unity Day - October 3 (1990)
Estimated 2010 population:	82,327,000
Population density:	597 persons per mi^2 (231 per km^2)
Population distribution:	74% urban, 26% rural
Life expectancy in years:	Male, 77; female, 82
Doctors per 1,000 people:	3.4
Birth rate per 1,000:	8
Death rate per 1,000:	10
Infant mortality:	4 deaths per 1,000 live births
Age structure:	0-14: 14%; 15-64: 66%; 65 and over: 20%
Internet users per 100 people:	75
Internet code:	.de
Language spoken:	German
Religions:	Protestant 34%, Roman Catholic 34%, Muslim 4%, other 28%
Currency:	Euro
Gross domestic product (GDP) in 2008:	$3.668 trillion U.S.
Real annual growth rate (2008):	1.3%
GDP per capita (2008):	$44,501 U.S.
Goods exported:	Chemicals, machinery, manufactured goods, motor vehicles
Goods imported:	Chemicals, food, machinery, petroleum and petroleum products
Trading partners:	Belgium, China, France, Italy, Netherlands, United Kingdom, United States

Germans are descended from a number of ancient tribes, including the Cimbri, Franks, Goths, and Teutons. A group of Slavic people called Sorbs live in the eastern part of the country. Most of the people living in Germany were born there. Most non-Germans who live in the country moved there with their families as guest workers. They came mostly from Turkey, the Balkan Peninsula, and Italy.

About one-third of Germans are Protestants, mostly Lutherans. Another third are Roman Catholics. About 4 percent are Muslims. Two main forms of the German language have long been spoken in Germany—High German in the south and center and Low German in the north. Today, a standardized form of the High German language is called Standard German.

Germany is a cultural, economic, and political center of Europe. It shares borders with the Netherlands, Belgium, Luxembourg, France, Switzerland, Austria, the Czech Republic, Poland, and Denmark.

HISTORY: BEFORE 1945

About 1000 B.C., warlike tribes began to migrate from northern Europe into what is now Germany. In the A.D. 400's, these Germanic tribes had pushed farther south and plundered the city of Rome. They eventually carved up the western portion of the Roman Empire into tribal kingdoms, of which the kingdom of the Franks became the largest and most important. In 800, the most famous Frankish ruler, Charlemagne, was crowned emperor of the Romans.

In 919, the Saxon *dynasty* (a series of rulers from the same family) came to power over the part of Charlemagne's empire that lay east of the Rhine River. It ruled until 1024. The third Saxon ruler, Otto I, extended the borders of his kingdom and, in 962, was crowned emperor in Rome. This marked the beginning of the Holy Roman Empire.

Although Frederick I—called *Barbarossa* or *Red Beard*—had further expanded the empire's borders by 1152, a string of generally weak rulers after the Saxons reduced the emperor's power, and by the 1300's the emperor was almost powerless. Beginning in 1438, the Habsburgs—one of Europe's most famous royal families—ruled the Holy Roman Empire almost continuously until 1806. The center of Habsburg power lay in what is now Austria.

1500–1648

In 1517, the writings of a German monk named Martin Luther sparked the Protestant Reformation, which quickly spread through Europe. Luther protested some practices of the Roman Catholic Church and called for reforms. By 1600, the German lands were divided by political and religious rivalries.

In 1618, a Protestant revolt set off a series of wars that lasted for 30 years. The Peace of Westphalia ended the Thirty Years' War in 1648. Under this treaty, Germany became a collection of free cities and hundreds of states.

1648–1945

In the mid-1600's, Frederick William of the Hohenzollern family began to add lands to his state in the Berlin area. In 1701, his son, Frederick I, was given the title king of Prussia. By 1763, Prussia, under Frederick II (the Great) was recognized as a great power.

The French Revolution and the Napoleonic wars caused many changes throughout Europe from 1789 to 1815. In

Adolf Hitler led the National Socialist (Nazi) Party. After the Nazis gained power in 1933, Hitler set up a brutal dictatorship and led Germany to defeat in World War II (1939-1945).

Wilhelm I was proclaimed the first *kaiser* (emperor) of the German Empire in 1871 in the Hall of Mirrors at the Palace of Versailles. Germany had just defeated France in the Franco-Prussian War. It was newly united under the leadership of the Prussian prime minister, Otto von Bismarck, who became Germany's first chancellor in 1871.

TIMELINE

Martin Luther (1483-1546)

Frederick II, the Great (1712-1786)

Otto von Bismarck (1815-1898)

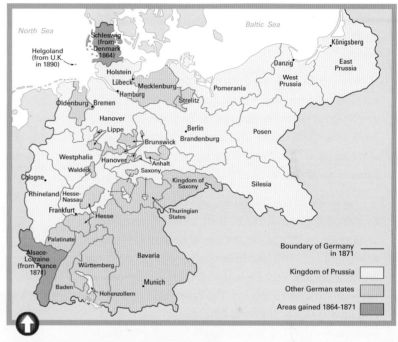

The Kingdom of Prussia dominated the political map of Germany by 1871. Other German states were located primarily in the southern part of the country.

1815, the powers that had defeated the French emperor Napoleon I met in the Congress of Vienna to restore order. Members of the Congress, including Prussia, divided Napoleon's lands among themselves. They also set up the German Confederation, a union of 39 independent states.

Economic and political unrest led many Germans to revolt against the Confederation, but the Revolution of 1848 was defeated. The German Confederation was reinstated in 1849.

In 1862, Otto von Bismarck became Prussia's prime minister. In 1871, following the Franco-Prussian War, Bismarck succeeded in uniting Germany. Wilhelm I was crowned the first *kaiser* (emperor) of the new German Empire in 1871, and Bismarck was made chancellor and head of government.

World War I began in 1914. Although Germany won a number of victories in the early years, the Allies defeated them in 1918. After the war, under the Treaty of Versailles, Germany lost its colonies and some of its European territory. Rejecting the government that had led them into the war, Germans created the Weimar Republic in 1919, but it was a weak state from the start.

During the political confusion of the 1920's and the early 1930's, the infamous Nazi Party rose to power. In 1933, Adolf Hitler was appointed chancellor of Germany. He proceeded to build a dictatorship, which he called the *Third Reich*, or *Third Empire*. Hitler was determined to assert German superiority over Jews and other non-Germanic peoples and to expand Germany's territory. His actions resulted in World War II (1939–1945).

During the war, the Nazis murdered about 6 million Jews and millions of others. However, the tide turned against Germany in 1943. Hitler committed suicide on April 30, 1945, just before Germany's surrender to the Allies on May 7.

HISTORY: 1945 TO PRESENT

World War II ended in Europe with Germany's unconditional surrender on May 7, 1945. In June, the Allied Big Four—France, the Soviet Union, the United Kingdom, and the United States—officially took control in Germany, dividing the country into four zones with each power occupying one zone. Berlin, located deep in the Soviet zone, was also divided into four sectors. Some lands in eastern Germany were lost to Poland and the Soviet Union.

Although the four powers had agreed to rebuild Germany as a democracy, it soon became clear that the Soviet Union intended to establish a Communist government in its zone. The tensions that subsequently arose between the Western powers and the Soviet Union came to be known as the Cold War.

France, the United Kingdom, and the United States combined the economies of their zones and prepared to unite them politically, but the Soviet Union kept its zone apart. In June 1948, the Soviets attempted to drive the Western powers from Berlin by blocking all highway, rail, and water routes to the city. The Allies retaliated by setting up a massive airlift that flew tons of supplies to Berlin every day. The Soviets finally lifted the blockade in May 1949. The airlift ended in September.

East and West divided

As a result of the political division between the Eastern and Western zones, Germany was divided into two states in 1949. The three Western zones became the Federal Republic of Germany (West Germany), with Bonn as its capital. The Soviet zone became the German Democratic Republic (East Germany), with East Berlin as its capital. Both states became officially independent from their occupying powers in 1955.

By that time, the amazing economic recovery of West Germany had helped the country achieve both prosperity and political stability. East Germany's economy also recovered under the leadership of Communist Walter Ulbricht, but its standard of living remained much lower

After the end of World War II in 1945, Germany was divided into four zones, controlled by the United States, France, Russia, and Great Britain. Germany lost much of its territory in the east to Poland.

than West Germany's. Every week, thousands fled to West Germany. Almost 3 million East Germans left, and the country's labor force fell dramatically.

Most refugees fled through Berlin, because the Communists had sealed off the East-West border in the 1950's. In August 1961, the East German government built a wall between East and West Berlin to prevent people from leaving East Germany. The Berlin Wall became a hated symbol of the divided Germany.

German reunification

In 1989, in response to public protests, Communist Hungary removed the barriers along its borders with non-Communist Austria. Immediately, thousands of East Germans fled through Hungary to West Germany, and citizens throughout East Germany protested for more freedom. In a dramatic policy change, the East German government responded by opening its own borders on November 9.

In March 1990, free elections took place in East Germany for the first time. Most East Germans voted for non-Communist candidates who favored rapid unification. Most West Germans also supported unification. In July, the economies of the two German states were united, and talks about unification

TIMELINE

May 7, 1945	Germany surrenders, ending World War II in Europe.
June 1945	The Allied powers divide Germany and the city of Berlin into four zones.
July-August 1945	Leaders of the United Kingdom, the Soviet Union, and the United States meet in Potsdam, where they agree to govern Germany together and to rebuild it as a democracy.
June 1948	Under the Marshall Plan, U.S. aid pours into the Western zone.
1948-1949	A Soviet blockade fails to force the Western Allies out of Berlin.
1949	East Germany and West Germany are established.
1955	East and West Germany are declared independent.
1961	The East German government erects the Berlin Wall to prevent East Germans from escaping to West Berlin.
1971	West German Chancellor Willy Brandt wins the Nobel Peace Prize for his efforts to reduce tensions between Communist and non-Communist nations. He works to normalize relations between East and West Germany.
1973	East and West Germany ratify a treaty calling for closer relations between the two nations. Both nations join the United Nations (UN).
1989	East German leader Erich Honecker is forced to resign. East Germany opens the Berlin Wall and other border barriers, allowing its citizens to travel freely to West Germany for the first time since World War II.
March 1990	East Germany holds free elections, resulting in the end of Communist rule there.
Oct. 1990	East and West Germany are unified and become a single nation.
Dec. 1990	Free elections make Helmut Kohl chancellor of the united country.
2002	The euro replaces the Deutsche mark.
2005	Angela Merkel becomes Germany's first woman chancellor.
2011	Angela Merkel plays a primary role in solving eurozone debt crisis.

Erich Honecker
(1912-1994)

Helmut Kohl
(1930-   )

Angela Merkel
(1954-   )

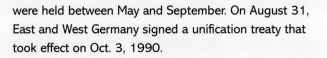

Fireworks illuminate the Brandenburg Gate in Berlin on Nov. 9, 2009, during celebrations that marked the 20th anniversary of the fall of the Berlin Wall.

were held between May and September. On August 31, East and West Germany signed a unification treaty that took effect on Oct. 3, 1990.

Germany's first national elections after unification took place on Dec. 2, 1990. Helmut Kohl, who had served as chancellor of West Germany from 1982 to 1990, became chancellor. He led a coalition of the Christian Democrats and Free Democrats.

Germany experienced a wave of social unrest in the early 1990's. Some neo-Nazis and other right-wing Germans began attacking immigrants, resulting in a number of deaths. Many Germans took part in demonstrations that protested the attacks.

After elections in 1998, Gerhard Schröder of the Social Democratic Party became chancellor. Angela Merkel of the Christian Democratic Union succeeded him in 2005. She continued as chancellor after her party received the most votes in parliamentary elections in 2009. Merkel became a strong leader among the nations who belong to the European Union (EU). When Europe experienced an economic crisis that threatened the EU's common currency—the euro—she was credited in 2011 with developing economic policy that helped the euro survive.

ENVIRONMENT

Germany is a land of scenic beauty with a varied landscape. The country has 16 states spread across five main land regions. The regions are the North German Plain, the Central Highlands, the South German Hills, the Black Forest, and the Bavarian Alps.

Northern Germany

The North German Plain—the largest land region— covers all of northern Germany. This low, flat plain slopes down toward the coastal areas of the North and Baltic seas. Broad, slow-moving rivers that flow northward into these seas drain the region. Many of these rivers, such as the Elbe, Ems, Oder, Rhine, and Weser, are important commercial waterways.

The wide river valleys of the North German Plain have soft, fertile soil, but large areas between the valleys are covered with sand and gravel deposited by glaciers thousands of years ago. However, the soil on the southern edge of the plain is highly fertile, and this area is heavily cultivated and densely populated.

Central and southern Germany

The Central Highlands, which include a mixture of landforms ranging from nearly flat land to mountainous areas, are covered with rock and poor soil. Peaks in the Harz Mountains and the Thuringian Forest rise more than 3,000 feet (910 meters).

The many rivers that flow through the Central Highlands have carved steep, narrow valleys in the region. These rugged gorges, especially that of the Rhine River, are among the most beautiful areas in Germany.

The South German Hills include a series of long, parallel ridges that extend from southwest to northeast. The lowlands between these ridges have fertile clay soil that provides some of the best farmland in Germany. Much of this region is drained by the Rhine River and two of its branches—the Main and Neckar rivers. The Danube River, which drains the southern part of the hills, is the only major river in Germany that flows eastward.

The Oker River has its source in the Harz Mountains on the northern edge of the Central Highlands. The climate in this region is marked by biting mountain winds, cool summers, and snowy winters.

The mountainous Black Forest region gets its name from the thick forests that cover its granite and sandstone hills. The Black Forest is known for its mineral springs and health resorts.

The varied landscape of Germany includes the flatlands of the North German Plain, the rugged hills of the Central Highlands, and the rocky ridges of the South German Hills. The country's major land regions also include the Black Forest, a mountainous region of southwestern Germany; and the Bavarian Alps, which form part of Germany's southern border. The North German Plain covers the largest land region in the country. Many of Germany's oldest cities, including Bonn and Cologne, are located in this area.

The island of Rügen, with its steep Stubbenkammer chalk cliffs, lies in the Baltic Sea off the North German Plain. Rügen is a popular resort area.

Lüneburger Heide, on the North German Plain, is one of the largest heathland areas in Germany. A portion of it was made Germany's first nature reserve in 1921. Largely infertile, the area supports sheep, forestry, and some farming.

The Black Forest, the setting for many old German fairy tales, is a rugged, mountainous region. Its name comes from the thick forests of dark fir and spruce trees that cover the mountainsides. The Feldberg, the highest peak in the Black Forest, rises 4,898 feet (1,493 meters). The region is also known for its mineral springs and health resorts.

The majestic, snow-capped Bavarian Alps, part of the largest mountain system in Europe, include the 9,721-foot (2,963-meter) Zugspitze, the highest peak in Germany. The Bavarian Alps region, with its sheer rock faces and crystal-clear, glacial lakes, is a favorite tourist spot.

Germany's mild climate is largely due to west winds from the sea that help warm the country in winter and cool it in summer. Southern areas located away from the sea have colder winters and warmer summers. Most of Germany receives moderate rainfall the year around, but deep snow covers some mountainous areas throughout the winter.

MUNICH

Munich is the capital of Bavaria, the largest of Germany's federal states. It is the third largest city in Germany, ranking in population after Berlin and Hamburg. The city is less than 100 miles (160 kilometers) from Brenner Pass, which straddles the border of Austria and Italy in the Alps. Its location, near where northern and southern Europe meet, has made Munich a major transportation link.

Munich was founded in 1158 by Duke Henry the Lion. In 1255, the city became the seat of a family called the House of Wittelsbach. This family ruled Munich and the rest of Bavaria until 1918.

Today, Munich is one of Germany's most important centers of economic activity. Electronics, publishing, and the production of chemicals are among the city's major industries. Munich is also well-known for its breweries.

Munich in the early 1900's

A small political group founded in Munich in 1919 became the Nazi Party in 1920. In 1923, the leader of this party, Adolf Hitler, proclaimed a Nazi revolution, or *putsch,* at a rally in a beer hall in Munich. The following day, Hitler tried to seize control of the Bavarian government, but he failed. This unsuccessful revolt became known as the "Beer Hall Putsch." However, Adolf Hitler became chancellor of Germany in 1933 and soon established a dictatorship. In 1938, the United Kingdom, Italy, and France—hoping to avoid war with Nazi Germany—signed an agreement at Munich to give Czechoslovakia's Sudetenland to Germany.

Historic and popular attractions

Although much of the city was destroyed during World War II, Munich's buildings were restored in their former style. As a result, many beautiful palaces, churches, and public buildings can be found throughout Munich. The Palace, the former residence of Bavarian monarchs, attracts visitors throughout the year. The German Museum houses

The twin towers of Munich's Cathedral rise above the rooftops and spires of the city's old section. Heavily damaged during World War II, this area has been carefully restored. The city's German name—München, or Place of the Monks—points to its origins as a monastic settlement established in the 700's. The city was founded in 1158 by Duke Henry the Lion.

Munich's Olympic Village, located northwest of the city center, hosted the 1972 Summer Olympics. The Olympic stadium, swimming pool, and sports hall could accommodate more than 100,000 people.

Munich's Oktoberfest runs for about 17 days from late September through early October. The festival, which dates back to 1810, attracts millions of visitors annually.

a world-famous exhibit of science and technology. And the twin towers of Munich's magnificent Cathedral, called the Frauenkirche, symbolize the city.

Three famous museums—the old Pinakothek, the new Pinakothek, and the Glyptothek—were restored after being severely damaged during the war. Some of the valuable paintings and sculptures in these museums were saved and are exhibited there today.

Munich is also a city of festivals and pleasure. In the summer, people get together in open-air beer gardens, or stroll in the many parks on the banks of the Isar River. Around the city's town hall in the heart of Munich, colorful street markets offer a variety of goods. And the Oktoberfest, the most famous beer festival in the world, is held in Munich from the middle of September to the beginning of October.

FRANKFURT

As its full name—Frankfurt am Main—implies, Frankfurt stands on the banks of the Main River. A network of railroads and highways links the city with all parts of Western Europe, and Frankfurt's airport is one of the largest and busiest on the continent. The city also ranks as one of Germany's busiest inland ports.

Frankfurt's role as the transportation hub of western Germany dates back to the time of the Roman Empire. The shallow part of the Main River where Frankfurt now stands provided the easiest north-south river crossing in all Germany, and it was widely used by travelers from all over Europe. The Franks who forded the river gave the city its name, which means *ford of the Franks.*

In 1798, the Rothschild family opened its first bank in Frankfurt, and the city is still a world center of commerce and banking. Frankfurt also hosts international trade fairs every year. Publishers and booksellers from around the world meet every autumn at the Frankfurt Book Fair. Consumer goods fairs are also held there regularly.

Before World War II, Frankfurt was Germany's best-preserved medieval city. During the war, however, Allied bombers leveled nearly half of the city, and today, gleaming skyscrapers have replaced some of Frankfurt's Gothic spires.

Historic sites

Although many of Frankfurt's buildings are modern, part of the old city still remains. The Römer—the city's ancient town hall—stands in the area surrounding Frankfurt Cathedral on the Main River, the historic center of the old town. Dating from the 1400's, this building contains the *Kaisersaal* (imperial room), the scene of the glittering celebrations and huge banquets that marked the coronations of the Holy Roman emperors. Nearby is St. Paul's Church, or Paulskirche, where leaders of the unsuccessful Revolution of 1848 met to draft a German national constitution. The great German writer Johann Wolfgang von Goethe was born in Frankfurt in 1749. His home has been reconstructed and is now a museum.

Frankfurt's city center was reduced to rubble as a result of World War II. Some of Frankfurt's medieval buildings have been reconstructed, but modern skyscrapers have replaced other old buildings.

The Frankfurt Opera House was reopened in the early 1980's—after extensive reconstruction—as a center for concerts and meetings.

Popular attractions

Although Frankfurt is one of western Germany's most lively cities, it retains a degree of small-town character. The Sachsenhausen district, for example, offers modern entertainment, but the narrow lanes that surround the area delight visitors with their Old World charm. Sachsenhausen is also the home of Germany's largest street market. The Schaumainkai, a riverside walkway, offers a fine view of the old town. A number of museums—Frankfurt's "museum mile"—line the walkway.

The city's traditional drink is a strong, bitter apple cider that goes especially well with *frankfurters*—the popular sausages to which the city lends its name. Experts believe that frankfurters were first made in Germany during the Middle Ages. In America, these sausages are called hot dogs. (Another favorite American food—the hamburger—is named after the German city of Hamburg.) Pork ribs with sauerkraut is also a hearty specialty of Frankfurt.

The characteristic gabled roofs of the Römer, Frankfurt's ancient town hall, contrast with the surrounding modern buildings.

The Frankfurt Book Fair is one of many international trade fairs held in the city each year.

HAMBURG AND BREMEN

In the late 1100's, the port cities of Hamburg and Bremen were important members of the Hanseatic League, a confederation of north German cities that grew out of trade associations. The Hanseatic League, whose members also included Lübeck and Cologne, controlled trade in the North and Baltic seas. Hamburg and Bremen, which both developed as a result of shipping and trade, remain thriving commercial and industrial centers today.

Hamburg

Hamburg—also known as "Germany's gateway to the world"—is located on the Elbe River, about 68 miles (110 kilometers) from the North Sea. This city-state ranks as Germany's most important port and the country's second largest city after Berlin. Hamburg is also an important political and cultural center.

Hamburg's industries include aircraft manufacturing, food production, ironworks, shipbuilding, and steelworks. The harbor, which stretches along the Elbe, is a center for foreign and inland shipping, and the hub of Hamburg's economic activities. Many of Germany's industrial products—including automobiles, chemicals, electronic goods, and machinery—are exported from Hamburg.

The city is also one of Germany's leading railroad centers and a major center for national publications. Such publications as the news magazine *Der Spiegel* and the newspaper *Die Zeit* are published in Hamburg. In addition, the city's many foreign consulates enable businesses in Hamburg to maintain contacts throughout the world.

Hamburg began to develop into an important trade center in the late 1100's. The city's leadership in the Hanseatic League in the 1200's considerably strengthened its position. In the late 1800's and early 1900's, Hamburg was a state in the German Empire and the Weimar Republic. The city-state's present constitution went into effect in 1952.

Hamburg harbor can accommodate large ocean-going vessels. The harbor connects Germany with other major trading nations throughout the world.

The magnificent facade of Bremen's town hall dates from the early 1400's. Bremen grew rich through its membership in the Hanseatic League, an association of trading cities.

On the embankment of the Weser River in Bremen stands old St. Martin's Church. The nearby Schnoor area of the city includes streets and buildings in the style of the 1400's and 1500's.

More than half of Hamburg's buildings were destroyed during World War II, along with large parts of the city's port and commercial areas. Since then, the city has been extensively rebuilt, and the *Rathaus* (town hall), the new opera house, and many other attractive buildings give Hamburg a modern appearance.

Bremen

The city-state of Bremen, the smallest of Germany's 16 federal states, consists of the cities of Bremen and Bremerhaven. Bremen, the state capital, lies along both banks of the Weser River, about 45 miles (72 kilometers) south of the North Sea.

While Bremen's economy is based chiefly on shipping and trade, other industries include shipbuilding, oil refining, food processing, and the production of automobiles, aircraft, electrical equipment, and textiles. Bremerhaven ranks as one of mainland Europe's largest fishing ports.

Bremen was founded sometime before A.D. 787, when it became a seat of bishops. The city gained economic importance in the 1300's through membership in the Hanseatic League. Like Hamburg, it became a state of the German Empire in the late 1800's. Bremen was also badly damaged during World War II, though the damaged areas were quickly rebuilt. Today, Bremen's landmarks include the Romanesque-Gothic Cathedral of St. Peter, begun in the 1000's, and the Gothic Rathaus, which dates from the early 1400's.

Fishermen in Bremerhaven sort through the day's catch. Bremerhaven is one of the greatest fishing ports of mainland Europe.

BERLIN

Berlin—now Germany's official capital and largest city—was also the capital of Germany before World War II. However, in 1949, four years after the war, Berlin—like Germany itself—was divided into two sections: West Berlin, which was allied with West Germany, and East Berlin, which became the capital of Communist East Germany. The city of Bonn was established as West Germany's capital.

In 1961, the Communists built a 26-mile- (42-kilometer-) long wall to stop East German refugees from fleeing to West Berlin. In 1989, after a movement for freedom, the Berlin Wall and other borders separating East and West Germany were opened. In 1990, East Germany and West Germany were unified. Although Bonn remained the seat of German government, a united Berlin became the official capital of the new Germany. It is also one of the country's 16 states.

The city

A new city grew up in Berlin after World War II. In West Berlin, areas devastated by the war were replaced with towering skyscrapers, wide streets, and large parks. Much of East Berlin was also rebuilt. Today, Berlin is one of Europe's most beautiful cities.

The Brandenburg Gate lies at the heart of downtown Berlin. The central section of the huge stone structure was built between 1788 and 1791. Today, the Brandenburg Gate is a famous symbol of the city. Unter den Linden (Under the Linden Trees), the grandest of prewar Berlin's avenues, runs east from the gate. Many cultural buildings line this street, including the State Opera House and Humboldt University. Marx-Engels Platz, once the site of mass rallies and demonstrations in East Berlin, lies at the end of Unter den Linden.

The Tiergarten (Animal Garden), a huge park that extends to the west of the Brandenburg Gate, includes the city's Zoological Gardens and Aquarium. Northwest of the Tiergarten stands the Hansa Quarter, which was designed by leading architects from a variety of nations. Southwest of the Tiergarten, fashionable stores and theaters line the Kurfürstendamm, one of Berlin's most famous boulevards. At its east end, the bomb-scarred tower of Kaiser Wilhelm Memorial Church stands as a reminder of war.

The Bode Museum is one of several museums that stand on Museum Island in Berlin. The museum is noted for its outstanding collections of sculpture, Byzantine art, and coins and medals. The building was completed in 1904.

Prenzlauer Berg is an area of Berlin known for its cafes and restaurants and for its historic buildings. The area is especially popular with young people who are attracted to the casual atmosphere.

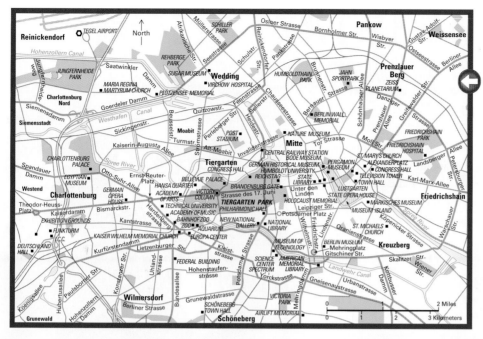

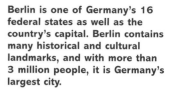

The Grunewald, an area of lakes and forests, stretches along the Havel River on the city's west side. The Olympic Stadium, built for the 1936 Olympic Games, stands just north of the Grunewald.

History before 1945

The village of Berlin grew up on the northeast bank of the Spree River in the 1200's. It was later united with the village of Kölln (or Cölln), founded on an island in the Spree. By the 1400's, Berlin was an important town in the province of Brandenburg. In 1470, the town became the official home of the Hohenzollern family, who ruled the province.

From 1640 to 1688, Berlin flourished under the rule of Frederick William of Hohenzollern. His son, Frederick I, became the first king of Prussia in 1701 and made Berlin his capital. In 1709, Berlin, Kölln, and three nearby settlements united as the city of Berlin.

During the 1700's, Berlin became a thriving trading and industrial center, and the arts and sciences also flourished. Immigrants from all over Europe flocked to the city. When the German Empire was formed in 1871, Berlin became its capital. The empire collapsed with Germany's defeat at the end of World War I in 1918. Berlin then became the capital of the weak Weimar Republic.

Berlin was especially hard hit by the worldwide economic depression of the 1930's. Hunger, unemployment, and widespread discontent paved the way for Adolf Hitler and his Nazi Party. In 1933, Hitler was elected chancellor of Germany, and in 1939, he initiated World War II.

By the end of the war in 1945, one-third of Berlin had been destroyed by bombings, and the city's population had dropped dramatically. On May 2, 1945, Berlin surrendered to the Allied Soviet army.

The glass dome of the Reichstag building provides a panoramic view of Berlin. Completed in 1999, the dome replaced the original 1894 dome, severly damaged in World War II bombings. The Reichstag houses Germany's parliament, the Bundestag.

ROOTS AND TRADITIONS

The German people are the descendants of various war-like tribes that migrated from northern Europe to what is now Germany sometime after 1000 B.C. In the 100's B.C., the tribes moved south to the Rhine and Danube rivers, the northern frontiers of the Roman Empire. The Romans named the tribes' land *Germania* and called the people *Germani,* though that was the name of only one tribe. Other tribes included the Cimbri, Franks, Goths, and Teutons.

The German language

Each of the early Germanic tribes had its own dialect. For many years, however, these existed only as spoken languages. The people composed ballads and stories about their gods and heroes and passed these on by word-of-mouth from one generation to the next.

The oldest known record of written German is a Latin-German dictionary dating back to around A.D. 770, the time of the Frankish ruler Charlemagne. Latin was the language used by the Romans, particularly among the nobles and the clergy. Charlemagne wanted his people to learn about the law and religion in a language similar to their own, and even encouraged German stories to be collected and written down. However, few works from this period survive. The German language commonly used today—Standard German—has its roots in Old High German, one of the dialects used in Charlemagne's time.

In general, there are two principal forms of spoken German. High German is spoken in the mountainous regions of central and southern Germany, and Low German is spoken in the lowlands of northern Germany. Regional dialects, however, still survive, and a dialect spoken in one region may not be understood by someone from another area.

Folk traditions

Many German folk traditions also differ from region to region. The people in the old mining towns around the Ore Mountains, for example, celebrate their particular history with processions, songs, and colorful costumes. In the state of Thuringia, the Thuringian Fair is still celebrated in the traditional style, and ancient wedding rituals are still observed.

A wedding procession provides an occasion for Sorbs to display their decorative costumes. This ethnic minority lives in the eastern part of the country. The people maintain their own Slavic language and traditions.

The May pole is a pagan symbol of May Day, a celebration of spring. In Bavaria, every village takes pride in its own "May tree."

A group of Bavarian women dress in traditional costumes to celebrate a Christian feast day.

The Pied Piper of Hamelin leads a procession of children in a summer commemoration of the famous legend. According to the story, the Pied Piper agreed to rid Hamelin (now Hameln) of its rats for a sum of money, but the mayor refused to pay him. Angry, the piper led the town's children into a cave. The cave closed upon them, and they were never seen again.

Some traditional folk festivals cross over regional boundaries, including many festivals originally held to honor the seasons and the harvest. The Shrove Tuesday festivals, for instance, were once rites of spring held in rural areas to greet the returning fertility of the earth. Seasonal traditions are also carried on in small villages and towns along the Rhine and Moselle rivers, where wine festivals celebrate the September grape harvest. The famous Oktoberfest held in Munich honors the harvesting of hops and other grains used to make beer.

As Christianity spread throughout Europe in the early Middle Ages, people began to mix their older customs with the new religious festivals. Thus, a pagan end of winter/birth of spring festival became Shrove Tuesday held before Lent, the 40-day period of spiritual renewal before Easter. Today, Shrove Tuesday is celebrated throughout Germany. The people of Cologne, Mainz, and other cities mark the occasion with wild merrymaking.

Participants in Shrove Tuesday festivals sometimes wear grotesque masks to "frighten away" the winter. To ensure a good harvest, masked dancers traditionally acted out the conflict between spring and winter.

ECONOMY

In 1945, at the end of World War II, Germany's economy lay almost in ruins. The controlling Allied powers worked to rebuild the economy, but the economic systems established in West Germany and Communist East Germany were entirely different.

The West German economy recovered at an amazing rate in the 1950's, greatly aided by the funds that the United States had begun to send under the Marshall Plan. By the mid-1960's, West Germany had one of the world's strongest economies. In East Germany, however, the Soviet Union set up a Communist state where the government controlled the production, distribution, and pricing of almost all goods. Under this system, East Germany grew to be one of the wealthiest Communist countries, but it lagged well behind West Germany.

When the two countries began economic unification on July 1, 1990, East Germany started to operate under West Germany's economic system. Goods that had been scarce in East Germany became readily available. However, East German industries and telecommunications systems had to be modernized. In addition, many businesses could not compete without government support and were forced to close, causing high rates of unemployment.

The changing economy posed challenges for the German government. In the former East Germany, people were disappointed at the slow rate of progress toward an improved economy. In the former West Germany, some people resented the cost of unification.

Industry

Manufacturing, Germany's fastest-growing industry, was the basis of West Germany's rapid economic recovery after World War II. Today, it remains central to Germany's economic strength. Germany has several major manufacturing regions, and there are factories almost everywhere. The Ruhr is the most important industrial region and one of the busiest in the world. It includes such manufacturing centers as Dortmund, Duisburg, and Düsseldorf. It produces most of the nation's iron and steel and has important chemical and textile industries.

Hanover's Exhibition Center hosts a major industrial fair each spring. The Hanover Fair, founded in 1947, ranks as the largest of its kind in the world and draws more than 6,000 German and foreign firms.

Much of Germany's steel is used to make automobiles and trucks, machinery, ships, and tools. The country also produces large quantities of cement, clothing, electrical equipment, and processed foods and metals. Other important products include cameras, chemicals, computers, leather goods, scientific instruments, toys, and paper.

Agriculture

German farmers produce only about two-thirds of the food consumed within the country. Germany is one of the world's largest importers of agricultural goods.

The chief grains grown by German farmers include barley, oats, rye, and wheat. Sugar beets, vegetables, apples, and other fruits are also important crops. Fine wines are made from grapes grown in vineyards along the Rhine and Moselle (or Mosel) rivers. Livestock and livestock products are also important sources of farm income.

Service industries

Industries that provide services rather than goods account for a large share of Germany's economic production. Finance, real estate, and business services form Germany's leading service industry group. Another important service industry group is community, government, and personal services. This group includes such activities as education, health care, advertising, car repair, and public administration. Other service industries involve trade, restaurants, hotels, transportation, and communication.

A Turkish vendor in Berlin finds plenty of customers for his oranges and bananas. Most non-Germans who live in the country moved there from Turkey, Yugoslavia, and Italy as Gastarbeitern (guest workers).

This map shows the economic uses of land in Germany. It also indicates the main farm and mineral products and identifies important manufacturing centers.

Potatoes have formed the basis of the German diet since the 1700's, when Frederick the Great ordered their cultivation. Today, potatoes rank among Germany's important crops.

HISTORIC SITES

The landmarks and artwork in many towns and cities throughout Germany reflect the country's long and rich heritage. Many of these historic sites were once the homes of great rulers. The western city of Aachen, for example, was the birthplace of Charlemagne, the most famous ruler of the Middle Ages. Charlemagne, who ruled in the late 700's and early 800's, made Aachen the capital of his western European empire. The city's magnificent cathedral was built during the emperor's rule and contains his tomb.

In 922, the founder of the Saxon dynasty, Henry I (the Fowler), built a fortress at Quedlinburg, at the foot of the Harz Mountains, as a stronghold against the invading Hungarians. Henry's son, Otto I, drove the Hungarians out of southern Germany in 955.

Nearby stand the Kyffhäuser Mountains. Frederick I—also known as *Barbarossa* or *Red Beard*—is said to "sleep" beside a huge table in a cave beneath the mountains. According to legend, Barbarossa, who was king of Germany from 1152 until 1190, will arise once his beard has grown completely around the table. Once risen, he will defeat Germany's enemies.

Sites of religious significance

Other cities in Germany are remembered for their religious significance. The Reformation, one of the most important religious movements in Europe, began in Wittenberg, a small university town in east-central Germany. The movement began when Martin Luther, a monk and professor of theology at the University of Wittenberg, attacked some of the church's practices.

On Oct. 31, 1517, Luther posted his famous Ninety-Five Theses on the door of Wittenberg's Castle church. As a result of this action, Luther was eventually expelled from the Roman Catholic Church and sentenced to death. However, Frederick the Wise, Prince of Saxony, protected Luther and concealed him at Wartburg, Frederick's castle in the Thuringian Forest.

Wartburg Castle crowns a steep hill overlooking the town of Eisenach. During the Middle Ages, a famous contest between wandering minstrels and poets took place in the castle. Martin Luther worked here in 1521 and 1522, translating the New Testament into German while under the protection of the Prince of Saxony.

Cities of culture

Many of Germany's cities and towns are also renowned cultural sites. Mainz, located at the junction of the Rhine and Main rivers, gained importance during the 700's when it became the seat of archbishops. However, Mainz is perhaps best known today as the home of Johannes Gutenberg, who invented the type mold that made printing from movable metallic type possible. Gutenberg used his invention to produce splendid books in Mainz during the 1400's.

The Chinese Teahouse stands on the grounds of Sans Souci park in Potsdam. Frederick the Great, the king of Prussia, started the park in 1745. The teahouse was completed in 1757 and now contains a fine collection of Chinese porcelain.

Charlemagne's marble throne sits in the cathedral at Aachen. Aachen was the Frankish emperor's birthplace and his capital during the late 700's and early 800's. The cathedral also holds Charlemagne's tomb.

Nuremberg stands on the Pegnitz River, which divides the central part of the city in half. During the late Middle Ages, Nuremberg became a prosperous trade and cultural city.

During the late 1400's and early 1500's, a number of German artists lived in the Bavarian city of Nuremberg. Albrecht Dürer, the most famous painter and printmaker in the history of German art, was born in Nuremberg in 1471. One of Dürer's most famous oil paintings, *Four Apostles,* was painted for the Nuremberg city hall in 1526. The sculptor Veit Stoss also settled in the city and produced several statues there in the late Gothic style of sculpture.

Weimar, once the capital of the grand duchies of Saxe-Weimar and Eisenach, also has a long connection with the arts. In the 1770's, after Duchess Anna Amalia founded the "court of the Muses" in Weimar, the city began attracting such giants of German culture as Johann Wolfgang von Goethe and Johann Christoph Friedrich von Schiller. In 1842, the Hungarian composer Franz Liszt was appointed court music conductor and made Weimar one of Europe's musical centers. The city also became a center for architecture in 1919 when the Bauhaus, an influential school of design, was founded there by Walter Gropius.

A short distance from Weimar lies Jena, another center of German intellectual life. In the late 1700's, Jena witnessed the birth of German Romanticism, a style of art and literature that stressed emotion over reason.

Frederick II of Prussia, known as Frederick the Great, was another ruler who was a benefactor of the arts, particularly architecture. During his reign, from 1740 to 1786, many of the most beautiful buildings in the royal seat of Potsdam were erected. Frederick himself planned the most famous of them all, the Palace of Sans Souci.

GHANA

The tropical country of Ghana lies on the Gulf of Guinea, where the continent of Africa bulges westward into the Atlantic Ocean. From its heavily populated coastal plain along the Gulf of Guinea, Ghana rises to the Kwahu Plateau. The plateau runs from the northwest to the southeast across the center of the country.

The Kwahu Plateau helps form a divide between Ghana's rivers. In the south and west, the Tano, Anko-bra, and Pra rivers flow south to the gulf. In the north and east, rivers such as the Black Volta and White Volta flow into Lake Volta in east-central Ghana. The main Volta River then flows from the lake to the gulf. Lake Volta, one of the world's largest artificially created lakes, was formed when the Akosombo Dam was built on the Volta River in 1965.

In general, Ghana has a tropical climate, with an average temperature of 80° F (27° C) in the south and warmer temperatures in the north. Most of Ghana receives 40 to 60 inches (100 to 150 centimeters) of rain a year, but heavier rains fall in the southwest. Axim, on the southwestern coast, receives more than 80 inches (200 centimeters) annually. Northern and eastern Ghana have a *savanna climate,* with severe dry spells from November to March.

Variations in Ghana's climate lead to differences in vegetation throughout the country. Southwestern Ghana is heavily forested. Farther north, the land gradually becomes *savanna* (thinly wooded grassland) that merges with grassy plains.

Tropical hardwoods such as mahogany are a valuable natural resource of the forests of Ghana. The country also has important mineral deposits of *bauxite* (used to make aluminum), diamonds, gold, and manganese. However, Ghana is largely an agricultural country, and its economy depends greatly on its farm products.

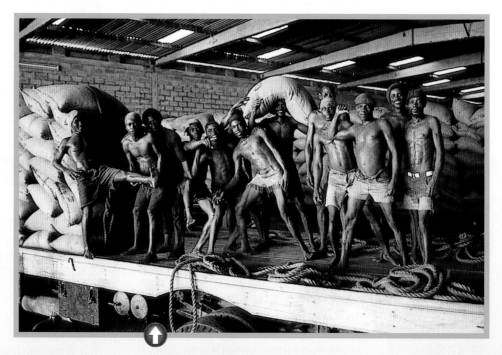

Cacao workers prepare sacks of cacao beans for export in a Ghana warehouse. Cacao seeds are the country's most important crop and leading export.

Moshi-Dagomba women sort rice, part of an abundant harvest in northern Ghana. Women raise food crops, while men grow cacao and other export crops on the country's small farms.

The White Volta River flows through northern Ghana into Lake Volta, one of the world's largest artificial lakes. The waters of this huge reservoir help produce electricity for much of the country.

An agricultural worker harvests the fruit of the oil palm, which is processed to produce palm oil. One of the most widely used vegetable oils in the world, palm oil is a basic ingredient in many kinds of soaps, ice creams, and margarines.

Cacao, a seed from which cocoa is made, is the most important crop and Ghana's chief export. Other valuable crops include coffee; coconuts; kola nuts, which are used to make soft drinks; and palm kernels, which are used to make oils.

Most of Ghana's factories are small plants that process agricultural products or the nation's timber. Manufactured goods include beverages, cement, and clothing. An aluminum smelter at Tema is the largest factory in the country. Hydroelectric power plants at the southern end of Lake Volta produce electricity for much of Ghana as well as the nearby countries of Togo and Benin.

During the 1970's and early 1980's, Ghana suffered severe economic problems. The economy of Ghana began to improve in the late 1980's, largely because of economic reforms. The reforms included a plan to grow and market crops other than cacao. Ghana had long depended on cacao exports for much of its income, and the nation's economy was seriously hurt when prices for cacao dropped. The addition of other crops, such as avocados, pineapples, and papayas, made Ghana less dependent on cacao. At the same time, the shortage of basic foods in local markets was eased when farmers started growing corn, sorghum, yams, and cassava.

The government also encouraged Ghanaians to catch and sell lobster and shrimp. In addition, it promoted the development of light industry, such as furniture manufacture. Also, foreign companies were encouraged to prospect in Ghana for gold and other minerals. New deposits of gold were soon discovered, abandoned mines were rebuilt, and gold production increased.

GHANA TODAY

When Portuguese explorers landed in what is now Ghana in 1471, they found so much gold there that they called it the Gold Coast. When the Gold Coast became an independent nation in 1957, it was named the Republic of Ghana after an ancient and powerful African kingdom.

Ghana was the first member of the British Commonwealth of Nations to be governed by black Africans. However, a series of military revolts since Ghana's independence often left the country in the hands of army officers.

Lieutenant Jerry Rawlings became Ghana's military head of state in 1981, when he overthrew the civilian government. In 1992, Ghana's voters approved a new constitution. Political parties were made legal and multiparty elections were held. Rawlings was elected president. His party, the National Democratic Congress (NDC), won a majority in parliament. Rawlings was reelected in 1996. In 2000, however, the opposition New Patriotic Party came to power, and John A. Kufuor became president. He was reelected in 2004. John Atta Mills of the NDC succeeded Kufuor in 2009.

The president of Ghana heads the government. The president appoints a cabinet to help carry out government functions. A 200-member parliament makes the country's laws. Ghana is divided into 10 regions for purposes of local government.

Almost all the people of Ghana—99 per cent—are black Africans. They belong to about 100 different ethnic groups, including the Ashanti, the Fante, the Ewe, the Ga, and the Moshi-Dagomba. The Ashanti and the Fante, who make up a large percentage of Ghana's population, are closely related and belong to an even larger cultural group of African peoples called the Akan. Many Ghanaians speak the African language of their ethnic group, but a large number also speak English, the official language of the country.

FACTS

Official name:	Republic of Ghana
Capital:	Accra
Terrain:	Mostly low plains with dissected plateau in south-central area
Area:	92,098 mi² (238,533 km²)
Climate:	Tropical; warm and comparatively dry along southeast coast; hot and humid in southwest; hot and dry in north
Main rivers:	White Volta, Black Volta, Ankobra, Pra, Tano
Highest elevation:	Unnamed location, 2,900 ft (884 m)
Lowest elevation:	Atlantic Ocean, sea level
Form of government:	Constitutional democracy
Head of state:	President
Head of government:	President
Administrative areas:	10 regions
Legislature:	Parliament with 230 members serving four-year terms
Court system:	Supreme Court
Armed forces:	13,500 troops
National holiday:	Independence Day - March 6 (1957)
Estimated 2010 population:	24,842,000
Population density:	270 persons per mi² (104 per km²)
Population distribution:	51% rural, 49% urban
Life expectancy in years:	Male, 58; female, 60
Doctors per 1,000 people:	0.2
Birth rate per 1,000:	30
Death rate per 1,000:	9
Infant mortality:	71 deaths per 1,000 live births
Age structure:	0-14: 38%; 15-64: 58%; 65 and over: 4%
Internet users per 100 people:	4
Internet code:	.gh
Languages spoken:	English, African languages (including Akan, Moshi-Dagomba, Ewe, Ga-Adangbe)
Religions:	Christian 68.8%, Muslim 15.9%, other 15.3%
Currency:	Cedi
Gross domestic product (GDP) in 2008:	$16.66 billion U.S.
Real annual growth rate (2008):	6.3%
GDP per capita (2008):	$707 U.S.
Goods exported:	Cocoa, diamonds, fruit and vegetables, gold, timber
Goods imported:	Machinery, motor vehicles, petroleum and petroleum products, rice
Trading partners:	China, Germany, Netherlands, Nigeria, South Africa, United Kingdom, United States

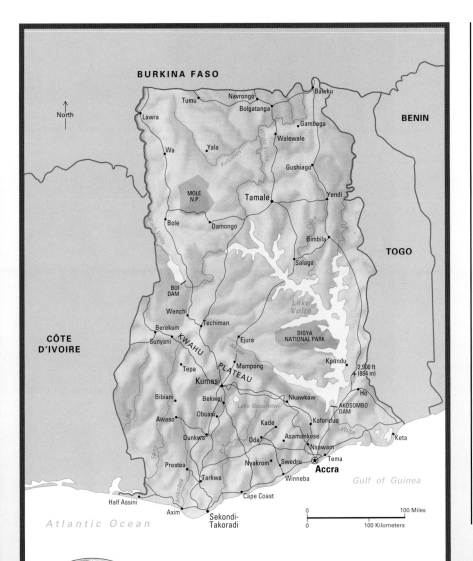

The Republic of Ghana lies on the Gulf of Guinea, a part of the Atlantic Ocean. Known in colonial times as the Gold Coast, it became an independent nation in 1957.

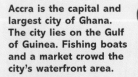

About half of Ghana's people live in rural areas and farm for a living. Many women raise food crops for their families on plots of ground, while men raise cacao on small farms. The cacao is sold to make chocolate and cocoa. Life in rural Ghana centers mainly on the village marketplace, where people come to buy and sell goods.

In central and southern Ghana, people live in rectangular houses with mud walls and thatched or tin roofs. Farther north, the mud houses are round with cone-shaped thatched roofs. Ghana's cities have many Western-style buildings, but large numbers of urban Ghanaians also live in traditional houses with mud-brick walls and tin roofs.

Traditional Ghanaian clothing is made from brightly colored cloth. Men wear it wrapped around their bodies, and women make it into blouses and skirts. But many Ghanaians wear Western-style clothing.

Few Ghanaians own an automobile, and so people often travel on crowded buses or flat-bed trucks. Some must walk from place to place.

Accra is the capital and largest city of Ghana. The city lies on the Gulf of Guinea. Fishing boats and a market crowd the city's waterfront area.

HISTORY

The earliest known inhabitants of the region that is now Ghana probably came from African kingdoms to the northwest during the 1200's. In the late 1600's, an Ashanti leader named Osei Tutu united his people and became the first Asantehene, or king, of this unified nation.

Tutu made the inland settlement of Kumasi the capital of his kingdom. During the 1700's, the Ashanti developed a powerful army and conquered surrounding territories. At its height in the early 1800's, the Ashanti Empire included much of modern-day Ghana, eastern Côte d'Ivoire, and western Togo.

Meanwhile, Europeans had come to the coast of Ghana. As early as 1471, Portuguese explorers landed on the shore and named the region the Gold Coast. Later, the Dutch came to compete with the Portuguese for gold. By 1642, the Dutch had seized all the Portuguese forts and ended Portuguese control of the coast.

During the 1600's, a large and profitable slave trade developed. The Dutch began competing with the English and the Danes for the trade. The buying and selling of slaves lasted for 200 years, ending in the 1860's. By 1872, the United Kingdom had gained control of the Dutch and Danish forts.

In 1874, the United Kingdom made the lands from the coast to the inland Ashanti Empire a British colony. The Ashanti and the British fought for control of trade in western Africa. In 1901, the British defeated the Ashanti and made their lands a colony.

The cacao industry prospered in the British colony in the early 1900's. The United Kingdom extended roads and railways, built hospitals, developed schools, and gradually gave the Africans more power. In 1946, a majority of the members of parliament in the colony were Africans elected to represent the people. However, most of the power was still in the hands of the British governor and cabinet.

Then in 1951, an African leader, Kwame Nkrumah, formed a cabinet, and in 1952 he became prime minister. By 1954, the people were running their own

The inner courtyard of a village in southern Ghana is surrounded by traditional, stoutly built mud-brick houses. In the late 1600's, the Ashanti people created an empire in this region that lasted until 1901.

government, except for police, defense, and foreign affairs. Finally, in 1957, full independence was granted, and the new nation of Ghana was born.

In 1960, the people of Ghana voted to become a republic, and they elected Nkrumah president. Through the mid-1960's, however, Nkrumah was committed to increasing his personal power. Government debt and corruption, along with the falling price of cacao, began to weaken the economy.

In 1966, a military council overthrew Nkrumah, suspended the Constitution, and dismissed the legislature. The council named General Joseph Ankrah head of the government. Ankrah resigned in 1969 and was replaced with Brigadier Akwasi Amankwa Afrifa.

A series of military revolts took place in Ghana during the 1970's. In September 1979, a civilian government took over, but in 1981, it too was overthrown. Lieutenant Jerry Rawlings then took control of the government and began economic reforms.

Before Rawlings seized power, many Ghanaians had moved to neighboring Nigeria to seek work. But Nigeria also began experiencing economic difficulties, and in 1983, it forced about 1 million people to return to Ghana. The return of these people created shortages of food, housing, water, and jobs in Ghana. The situation led to tension between Nigeria and Ghana.

From 1983 to 1988, Ghana's government took strong measures to improve the economy. It tripled prices paid to cacao growers, which led to an increase in production. It also removed price controls. As a result of these moves, Ghana's economy experienced significant growth.

In 1992, Ghana's voters approved a new constitution. Political parties, which had been banned since 1981, were legalized. In multiparty elections held later that year, Rawlings was elected president, and Rawlings's party, the National Democratic Congress (NDC), won a majority of the seats in parliament.

In December 2000, John A. Kufuor of the New Patriotic Party was elected president. John Atta Mills of the NDC succeeded Kufuor in 2009.

Dressed in brightly colored robes, Ghanaian women gather for a ceremony. About 100 different ethnic groups live in Ghana today. Many are descendants of people from African kingdoms to the northwest of present-day Ghana who migrated to the region in the 1200's. The Ashanti probably are descended from people who lived in western Africa thousands of years ago.

A fortress on the coast of Ghana stands as a grim reminder of the African slave trade. From the 1600's to the 1800's, European slave traders built such forts as bases for their operations.

KWAME NKRUMAH

1909 Born in Nkroful, Gold Coast.
1935 Enters Lincoln University, United States.
1946 Publishes Towards Colonial Freedom.
1947 Returns to the Gold Coast.
1950 Imprisoned by British.
1951 Wins the Gold Coast's first general election.
1957 Sees birth of the nation of Ghana.
1958 Legalizes imprisonment without trial.
1961 Presides over widespread unrest.
1962 Survives assassination attempt.
1964 Becomes president for life of one-party state.
1966 Is overthrown by army.
1967 Exiled in Guinea.
1972 Dies in Conakry, Guinea.

The independence movement that swept through the countries of Africa after World War II (1939–1945) owes much to the work of Kwame Nkrumah. In the 1940's, Nkrumah sought self-government for his country, then called the Gold Coast. In 1951, the British, who controlled the country, asked him to form a cabinet, and in 1952 he became prime minister. He continued to lead the Gold Coast's drive for full independence. Finally, in 1957, the Gold Coast became the first black African colony to win its freedom. In 1960, Nkrumah was elected president. In an effort to develop Ghana's economy and improve living conditions, he promoted industry, introduced health and welfare programs, and expanded the school system. But he also began jailing his opponents, and his government became corrupt. The economy began to weaken, and in 1966, when Nkrumah was visiting China, army leaders took control of the country. Nkrumah went into exile in nearby Guinea, where Guinean President Sékou Touré made him honorary president. He died there in 1972.

GIBRALTAR

The tiny peninsula of Gibraltar, a British territory at the southern tip of Spain, has a land area of only 2.5 square miles (6.5 square kilometers), but its landmark—the great Rock of Gibraltar—is recognized the world over. For many people, this huge limestone mass has come to symbolize strength and security.

The Rock of Gibraltar lies at the entrance to the Mediterranean Sea, rising 1,398 feet (426 meters) above sea level on the north and east side. From the top of the Rock on a clear day, visitors can see as far as Spain's Sierra Nevada and Morocco's Atlas and Rif mountains.

For centuries, the Rock of Gibraltar has been the subject of myth and legend. In Greek mythology, it was one of the Pillars of Hercules. According to legend, Hercules created the Rock by tearing a mountain in two, thus separating Europe and Africa and creating the Strait of Gibraltar.

Early conquests

The ancient Phoenicians were the first to establish a trading post on the Rock, realizing that the location of Gibraltar gave it great strategic importance. The Moors of North Africa settled at Gibraltar in A.D. 711 and held it for almost 600 years. Their leader, Tariq ibn Ziyad, ordered a fortress built high on the hillside overlooking the Bay of Gibraltar. Spain conquered Gibraltar in 1309 but lost it to the Moors again in 1333. Spain reclaimed ownership of Gibraltar in 1462 and held it until 1704 when it was captured by a British naval force, aided by the Dutch.

British base

The Rock was promptly developed as Great Britain's (later the United Kingdom's) major west Mediterranean base.

British naval power on Gibraltar played an important role in stopping Napoleon from conquering all of Europe during the 1800's. Later, during World War II (1939–1945), the Allies launched an attack from Gibraltar against German and Italian forces in North Africa. After the war, Gibraltar's military importance

A Barbary ape suns itself on an upper ledge of the Rock. The only wild monkeys in Europe, Barbary apes are about half the size of a large dog.

Fishermen prepare their nets on a beach in the shadow of the Rock of Gibraltar. The population of Gibraltar is largely made up of British military personnel and civilians descended from Italian, Maltese, Portuguese, and Spanish settlers. Many Moroccans now live and work on Gibraltar.

A ship awaits repair in dry dock at Gibraltar's main shipyard. Gibraltar has been an important British naval base since the early 1700's. Its extensive port facilities contribute much to the economy, but tourism is important as well.

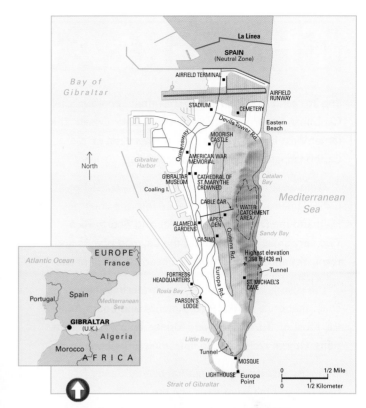

The Rock of Gibraltar covers most of its land area. Gibraltar has always had great military value, keeping enemy ships from entering or leaving the Mediterranean. The Strait of Gibraltar is about 32 miles (51 kilometers) long and 8 to 23 miles (13 to 37 kilometers) wide.

Metal sheeting on Summit Ridge, the Rock's highest point, drains rain water into catchment tanks. The tanks serve all of Gibraltar, which has virtually no water supply.

gradually declined. In 1991, the United Kingdom withdrew its military forces from Gibraltar.

In 1964, the United Kingdom considered granting independence to Gibraltar. Spain strongly objected, however. According to the terms of the Treaty of Utrecht, signed by Spain and Britain in 1713, Britain was obligated to offer Gibraltar to Spain if Britain ever decided to give up the dependency.

In 1967, after the people of Gibraltar voted for continued British control, the United Kingdom decided to keep the dependency. The Spanish government responded by closing the border, thus restricting overland travel between Spain and Gibraltar. The border was not fully reopened until 1985.

Vacationers and residents

In addition to being an important naval and air base, Gibraltar has also developed a thriving tourist trade. Its duty-free status offers vacationers great bargains on European luxury goods and North African craft products.

The Rock's most famous natives may well be the shaggy-haired creatures known as Barbary apes. These animals are actually tailless monkeys related to the rhesus monkey of India, and not true apes.

Protected by the British government, the Barbary apes wander freely over the upper Rock. Legend has it that the monkeys once warned the British of a surprise attack on Gibraltar by Spain, and that the British will never lose control of the Rock as long as the apes live there.

About 29,000 people live in Gibraltar, representing several ethnic groups. They include British, Italian, Maltese, Portuguese, Spanish, and Moroccans. Most people live in the town of Gibraltar. They work for the government, a dockyard, or in the tourist industry.

The West Indies are an island chain that separates the Caribbean Sea from the rest of the Atlantic Ocean. The larger West Indies, lying on the northern rim of the Caribbean, are called the Greater Antilles.

From west to east, the Greater Antilles consist of Cuba, Jamaica, Hispaniola (divided into the countries Haiti and the Dominican Republic), and Puerto Rico. Two very small island groups, the Cayman Islands and the Turks and Caicos Islands, lie near the Antilles area.

The islands are part of an underwater mountain chain that once linked North America and South America. Most of the islands were formed by volcanic eruptions. Others are coral and limestone formations.

The islands have great natural beauty. Sandy beaches and tall palm trees line the coasts, while lush tropical vegetation covers many of the islands. The many varieties of flowering plants on the islands include bougainvillea, hibiscus, orchid, and poinsettia. Such sport fish as marlin and sailfish, as well as brightly colored tropical fish, swim in the blue-green waters.

The islands have a warm, tropical climate, with temperatures averaging 80° F (27° C) in the summer and 75° F (24° C) in the winter. Rainfall averages 60 inches (150 centimeters) a year, with some mountainous areas receiving up to 200 inches (500 centimeters). Hurricanes often strike the islands, chiefly during the late summer and early fall.

The first inhabitants of the Greater Antilles were Carib Indians. In 1492, Christopher Columbus became the first European to see the islands. Various European nations, especially Great Britain, Spain, and France, eventually gained control of different islands. Today, most of the islands are independent nations.

Cayman Islands

The Cayman Islands are a British dependency about 200 miles (320 kilometers) northwest of Jamaica in the Caribbean Sea. Three islands form the group— Grand Cayman, Little Cayman, and Cayman Brac. The capital and largest city, George Town, stands on Grand Cayman, the largest island.

Stingray City attracts swimmers, boaters, and snorkelers. The city gets its name from the tame sting rays that gather there in search of food handed out by the visitors.

FACTS

CAYMAN ISLANDS

Capital:	George Town
Islands:	Grand Cayman, Little Cayman, Cayman Brac
Area:	101 mi² (262 km²)
Form of government:	British dependency
Head of state:	British monarch
Estimated 2010 population:	50,000
Official language:	English
Religion:	Mainly Christian
Currency:	Caymanian dollar

Farm production is low on the islands, and most food must be imported. Taxes are extremely low in the Caymans, so many foreign companies conduct business there. These businesses and the tourist industry are important to the economy. The islands are popular with scuba divers. The Pirate's Week festival in October draws tourists who come to see costumes, parades, treasure hunts, and a staged pirate raid.

The Turks and Caicos Islands

The Turks and Caicos Islands are barren, sandy islands about 90 miles (145 kilometers) north of the Dominican Republic. Like the Caymans, the two island groups are a British dependency in the Commonwealth of Nations. The capital and largest city is Grand Turk on the island of Grand Turk. Many of the islanders make their living by fishing.

In 1512, the Spanish explorer Juan Ponce de León sighted the islands. According to legend, the Turks Islands got their name because a red cactus flower on the islands reminded an early settler of a Turkish fez. The Caicos Islands probably got their name from the Spanish word for a *cay,* or small island.

The most important industries on the islands are financial services, fishing, and tourism. Lobster is the chief export.

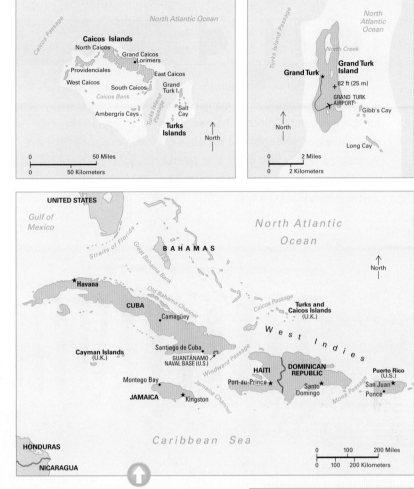

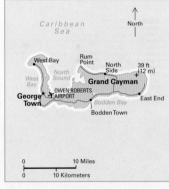

The Cayman Islands and the Turks and Caicos Islands lie near the Greater Antilles area. The Cayman Islands lie southwest of Cuba. The Turks and Caicos Islands lie north of the Dominican Republic. Both sets of islands are administered by the United Kingdom.

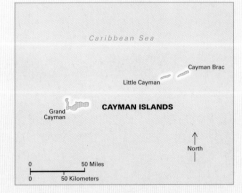

FACTS

TURKS AND CAICOS ISLANDS

Capital:	Grand Turk
Main islands:	Grand Turk, Salt Cay, South Caicos, East Caicos, Grand Caicos, North Caicos, Providenciales, West Caicos
Area:	166 mi² (430 km²)
Form of government:	British dependency
Head of state:	British monarch, represented by governor general
Estimated 2010 population:	24,000
Official language:	English
Religion:	Mainly Protestant
Currency:	United States dollar

GREECE

Greece is a mountainous country with a landscape that ranges from the sunny seacoasts of its many offshore islands to the rocky hills of the mainland. Greece's charming villages, where sea breezes turn windmills and whitewashed cottages glisten in the sunlight, have fascinated artists and poets throughout the ages. Even the rocky soil that covers most of the land sparked the imaginations of its ancient inhabitants.

According to an age-old Greek legend, god made the world by sifting earth through a strainer. He made one country after another with the good soil that sifted through. After he was finished, he threw away the stones that were left, and these stones became Greece.

Greece is a nation with a long and fascinating history, for on this nation's ancient shores, Western civilization began about 2,500 years ago. It was here that the ancient Greeks established the ideals of democracy. In the ancient monuments that dot the landscape, we see the traditions of justice, individual freedom, and representational government carved in stone.

The Greek people have received little benefit from the principles of democracy founded so long ago in their homeland. Shaky governments and political confusion have troubled Greece through most of its modern history. At times, strong military leadership has imposed control over chaos, but often at the expense of individual rights.

In the 500 years of Turkish occupation that began in the 1300's, the Greek people suffered terrible blows to their pride and sense of independence. Since it gained independence from Turkish control in 1829, Greece has rarely been free from political and economic difficulties. In 1897, Greece found itself at war with the Ottoman Empire over Turkish-held Crete. When Greece was declared a republic in 1923, its people were divided between the *republicans,* who supported the republic, and the *royalists,* who wanted a king. Later, the effects of World War II (1939–1945) almost destroyed the country's economy.

Greece adopted a new constitution in 1975. The document officially eliminated the monarchy that had ruled Greece, and it made the nation a parliamentary republic. The country's economy has expanded significantly, largely as a result of government programs, economic aid from the United States, and trade with the Middle East and with members of the European Union. Still, many families have lived in extreme poverty for generations.

But high on a hill in Athens, one of the magnificent achievements of the ancient Greeks can still be seen—the Acropolis. It stands as a reminder of the glories of the past and a symbol of hope for the future.

GREECE TODAY

The 1950's provided Greece with some relief from the political and financial problems it had suffered since the early days of independence. The Western powers provided massive military and economic aid to the country, and Greece joined the North Atlantic Treaty Organization (NATO) in 1952. During this time, the government improved the country's finances, controlled rising prices, and encouraged agriculture and industry to expand.

Rule of the colonels

In 1963, George Papandreou of the Center Union Party became prime minister of Greece. But Papandreou disagreed with King Constantine II on who should have political power and control over the armed forces. After Constantine dismissed Papandreou in 1965, another period of political confusion rocked the government of Greece. Only a month before new elections were to be held, the government was taken over by Greek army units.

Although King Constantine II remained head of state, he became powerless. A junta of three army officers led by Colonel George Papadopoulos set up a military dictatorship that suspended the rights of the people, prohibited all political activity, and made mass arrests. The junta also imposed harsh controls on newspapers, and it dissolved hundreds of private organizations of which it disapproved.

In 1973, Papadopoulos announced the end of the monarchy and declared that Greece would be a republic. But even as the government was preparing to hold parliamentary elections, a group of military officers once again overthrew the government.

In 1974, when Greek officers on the island of Cyprus helped Cypriot troops overthrow their government, Turkey accused Greece of violating the independence of Cyprus. Turkish troops then invaded the island. After several days of fighting, a cease-fire was signed.

Shortly after the cease-fire, the government of Greece collapsed under the combined pressure of the Cyprus crisis and economic recessions. Constantine Caramanlis, head of the New Democracy Party, then became prime minister.

FACTS

Official name:	Elliniki Dimokratia (Hellenic Republic)
Capital:	Athens
Terrain:	Mostly mountains with ranges extending into the sea as peninsulas or chains of islands
Area:	50,949 mi^2 (131,957 km^2)
Climate:	Temperate; mild, wet winters; hot, dry summers
Main rivers:	Vardar, Aliakmon, Pinios
Highest elevation:	Mount Olympus, 9,570 ft (2,917 m)
Lowest elevation:	Sea level along the coasts
Form of government:	Parliamentary republic
Head of state:	President
Head of government:	Prime minister
Administrative areas:	51 nomoi (prefectures), 1 autonomous region
Legislature:	Vouli ton Ellinon (Parliament) with 300 members serving four-year terms
Court system:	Supreme Judicial Court
Armed forces:	156,600 troops
National holiday:	Independence Day - March 25 (1821)
Estimated 2010 population:	11,197,000
Population density:	220 persons per mi^2 (85 per km^2)
Population distribution:	60% urban, 40% rural
Life expectancy in years:	Male, 77; female, 82
Doctors per 1,000 people:	5
Birth rate per 1,000:	10
Death rate per 1,000:	10
Infant mortality:	4 deaths per 1,000 live births
Age structure:	0-14: 14%; 15-64: 67%; 65 and over: 19%
Internet users per 100 people:	33
Internet code:	.gr
Languages spoken:	Greek (official), English, French, Turkish
Religions:	Greek Orthodox 98%, Muslim 1.3%, other 0.7%
Currency:	Euro
Gross domestic product (GDP) in 2008:	$357.55 billion U.S.
Real annual growth rate (2008):	2.8%
GDP per capita (2008):	$32,131 U.S.
Goods exported:	Aluminum, chemicals, clothing, fruit and vegetables, machinery, olive oil, petroleum products
Goods imported:	Crude oil, machinery, motor vehicles, pharmaceuticals, ships
Trading partners:	France, Germany, Italy, Netherlands, Russia, United Kingdom

BULGARIA

MACEDONIA

TURKEY

ALBANIA

Orestias
Dhidhimotikhon
Souflion
FALAKRON MTS.
Kaimakchalan
8,281 ft (2,524 m)
Drama
Xanthi
Komotini
Ferrai
Polikastron
Kilkis
Serrai
Dhoxaton
Kavalla
Khrisoupolis
Alexandroupolis
Florina
Edhessa
Nigrita
Struma
Thasos
Amindaion
Naousa
Yiannitsa
Langadhas
Thasos
Kastoria
Ptolemais
Veroia
Alexandria
Thessaloniki
Ierisos
Samothrace
Smolikas
8,652 ft (2,637 m)
Kozani
Servia
Katerini
KHALKIDHIKI
PENINSULA
Kariai
Mount Athos
Limnos
Konitsa
Siatista
Grevena
OLYMPUS N.P.
Mount Olympus
9,570 ft (2,917 m)
Litokhoron
Ouranoupoli
Dafni
Sikia
Mt. Athos
6,670 ft
(2,033 m)
Moudhros
VIKOS-AOOS N.P.
PINDOS N.P.
Elasson
Kassandra
Sithonia
Karousadhes
Ioannina
Kalabaka
Tirnavos
Ayios
Evstratios
Kerkira
Corfu
Igoumenitsa
Parga
Trikala
Larisa
Lesbos
Ayia Paraskevi
Paxoi
Kardhitsa
Velestinon
Volos
Pelagos
Yioura
Eressos
Mitilini
Arta
Perveza
Farsala
Almiros
Skiathos
Iliodhromia
Polikhnitos
Plomarion
TURKEY
Levkas
Timfristos
7,595 ft (2,315 m)
Gulf
of Volos
Skopelos
Northern
Sporades
Leucas
Agrinion
Lamia
Skiros
Skiros
Psara
Chios
Vrondadhes
Ithaca
Astakos
Thermon
MT. OETA N.P.
N. Gulf
of Euboea
Kimi
Khios
Aitolikon
Amfiklia
PARNASSOS N.P.
Parnassos
8,061 ft (2,457 m)
Euboea
Aegean
Sea
Cephalonia
Lixourion
Mesolongion
Amfissa
DELPHI
Dhelfoi
Levadhia
Khalkis
Aliverion
Argostolion
AINOS
N.P.
Gulf of Patrai
Navpaktos
Aiyion
Kato Akhaia
PARNITHA N.P.
Thebes
Marathon
Karistos
Andros
Killini
Erimanthos
7,297 ft (2,224 m)
Patrai
Kalavrita
Sikiona
Elevsis
Akharnai
★**Athens**
Neon
Karlovasion
Samos
Zakinthos
Lambia
PELOPONNESUS
Corinth
Corinth
Canal
Megara
Piraeus
Glifadha
SOUNION N.P.
Andros
Ikaria
Ayios
Kirikos
Akritis
Samos
Zante
OLYMPIA
MYCENAE
TIRYNS
Argos
Navplion
Aiyina
Aiyina
Lavrion
Yiaros
Tinos
Mikonos
Delos
Patmos
Lipso
Leros
Piros
Tripolis
Kranidhion
Kea
Siros
Kithnos
Siros
Kalimnos
Kalimnos
Kiparissia
Filiatra
PARNON MTS.
Idhra
Serifos
Paros
Naxos
Cos
Kos
Strofadhes
Gargalianoi
Kalamata
Sparta
Evrotas
Sifnos
Paros
Naxos
Amorgos
Pylos
Koroni
EPIDAURUS
LIMERAS
Milos
Cyclades
Astipalaia
Nisiros
Simi
Rhodes
Methoni
Cape Akritas
Gulf of
Messini
Yithion
Monemvasia
Milos
Sikinos
Ios
Folegandros
Anafi
Dodecanese
Sirna
Tilos
Khalki
Rhodes
Lindos
Areopolis
Neapolis
Cape Malea
Thira
Saria
Cape Matapan
Kithira
Cythera
Sea of Crete
Karpathos
Andikithira
Cape Spatha
Gulf of
Khania
Nea
Alikarnassos
Cape
Sidheros
Karapathos
Mediterranean
Sea
Kastellion
Khania
Rethimnon
Iraklion
KNOSSOS
Gulf of
Marabello
Sitia
Kasos
SAMARIA N.P.
Mt. Ida
8,058 ft (2,456 m)
Crete
Ierapetra
Ayios Nikolaos
Cape Krios
Palaiokhora
Timbakion
Koufonision
Cape Lithinon
PHAISTOS
Khrisi
Gavdhos

GREECE

North

100 Miles
100 Kilometers

The mainland of Greece lies on the southern tip of the Balkan Peninsula. The large island of Crete and 437 small islands lie in the Mediterranean Sea. The small islands make up a fifth of Greece's land area.

Evzones, or guards, dressed in traditional Greek uniforms, march before the Tomb of the Unknown Soldier at the Greek Parliament Building on Syntagma Square in Athens.

The restoration of democracy

In 1974, Greece held its first free elections in 10 years, and Greeks voted to make the country a republic. A new constitution was adopted in 1975, and civilian control over the military was established.

In 1981, Greece joined the European Community, which later became part of the European Union (EU). Also in 1981, the Panhellenic Socialist Movement (PASOK) party formed the first Socialist government in Greece. The Socialists were defeated in 1989 by the New Democracy Party (NDP) but won control again in 1993. PASOK governed Greece from 1993 to 2004 and again from 2009 to 2011. The NDP governed from 2004 to 2009.

In 2009, Greece's budget developed a severe deficit. By 2010, its financial problems threatened the stability of the euro, the EU's common currency. Greece accepted emergency loans from the EU in 2010 and 2011, but the loans required the Greek government to implement harsh austerity measures. In parliamentary elections in 2012, Greeks expressed their anger over the forced austerity by voting against PASOK and the NDP, the two main parties.

ENVIRONMENT

About 70 percent of the land in Greece is composed of limestone mountains and hills, which are either bare or covered with patches of thorny, woody shrubs. Yet, however desolate the upland areas may appear, they have a strange, rugged beauty all their own.

Cypress, fir, myrtle, and fig trees grow on the mountainsides. In the spring, blooming red poppies blanket the slopes. The uplands give way to river valleys, narrow coastal plains, and the long fingers of land that jut out into the Mediterranean Sea. These features give the Greek landscape its unique character.

Land regions

Greece has often been described as the "land of the mountains and the sea"—and, except for a few districts in Thessaly, no part of the country is more than 85 miles (137 kilometers) from the ocean. The mountain ranges and the sea divide Greece into several land regions.

The Pindus Mountains, which extend southward down the "backbone" of Greece, are an extension of the Dinaric Alps. The Pindus are composed mainly of limestone, with large areas of *karst* (eroded limestone). Many underground streams run through this striking landscape of steep slopes and deep ravines. Sheep and goats graze on the mountain pastures, while the region's two river valleys—the Ptolemais and the Ioannina—are the main population centers.

The coastal plains and lowlands are the center of the nation's agricultural development. The region of Thessaly, surrounded by tall mountains in the east-central part of the country, is known as the "breadbasket of Greece," because fields of wheat cover most of its cultivated land.

In northeastern Greece, the river basins and alluvial plains of Macedonia-Thrace produce plentiful harvests of tobacco and other crops. Farmers on the Salonika Plain, located at the southwestern tip of Macedonia-Thrace, grow cotton, fruits, rice, and wheat.

Ruins cling to the hillsides in what was once the town of Mistras on the Peloponnesian Peninsula. Founded as a Frankish fortress in 1248, the town was taken by Byzantine rulers in 1262. Under their control, Mistras became a center of Byzantine culture.

The Peloponnesus is a large peninsula connected to the Greek mainland by an isthmus. The Corinth Canal cuts across the isthmus, and virtually makes the Peloponnesus an island. This region of rugged mountains, small valleys, and rocky coasts is not suitable for farming, but the Peloponnesus is rich in history. The ruins of Corinth, Olympia, and other historic sites still stand on the peninsula.

The most populated region of Greece is the Southeastern Uplands, a region of mountains and hills with many small valleys among them. Athens, Greece's capital and largest city, is located on the southern tip of the Southeastern Uplands.

Climate and vegetation

Greece enjoys a typical Mediterranean climate, with hot, dry summers followed by mild, wet winters. However, the climate varies greatly, depending on altitude and location. Winters can be extremely cold and summers can be very hot in the north and inland regions, with snow and freezing temperatures in the mountains.

The trees, shrubs, and small plants that grow in Greece are well adapted to the long, dry summers. Such typical Mediterranean crops as lemons and olives thrive in the lower altitudes, and the air is scented by the aromatic herbs typical of the Mediterranean scrub called *maquis*. Another classic feature of the Greek landscape is the cypress tree.

Greece has suffered greatly from the loss of the dense forests of oak, chestnut, beech, and plane trees that once covered the land. The destruction of the forests began over 2,000 years ago, when the Greeks cut down trees for fuel and shipbuilding. During the 1900's, forestland was cleared for livestock grazing. With financial aid from other countries, the Greek government has now begun to take steps to reforest the land.

Mouse Island lies off the east coast of Corcyra (Kerkira or Corfu). According to legend, the island, called Pontikonísi in Greek, was once a boat that was turned to stone by Poseidon, the Greek god of the sea, after it carried Odysseus back to Ithaki (Ithaca).

Vineyards and olive groves form a pattern of cultivation in the valleys and plains that lie among the mountains on the island of Crete. Irrigation must be used on the island's many farms during the dry summer months.

Time has changed the rugged Greek coastline. The blue waters of the Mediterranean Sea splash against the rocks, but the natural beauty of Greece has attracted throngs of tourists and lined its beaches with hotels and resorts.

PEOPLE

About 98 percent of the people in Greece are ethnic Greeks. Minorities such as Macedonians, Turks, Albanians, and Bulgarians make up the rest. Almost all of the people in Greece speak modern Greek, a language that developed from classical Greek. Most of them belong to the Greek Orthodox Church.

Over the centuries, the Greeks have developed a strong cultural unity and a deep sense of national identity. This feeling of community has flourished in spite of the fact that the geography of Greece—a land of mountains divided by valleys and plains—has made communication difficult.

Waves of emigration

In recent years, many rural people have left their villages, hoping to find jobs and a better life in the cities. Today, about 40 percent of the people live in rural communities. Almost one-third of the entire population of Greece lives in the Athens metropolitan area.

Many Greeks have left their homeland to live in other countries. Some fled to escape the country's political turmoil, and others left to seek better jobs. After civil war broke out in Greece in 1945, political refugees were forced to flee the country. In the early 1960's, many Greek workers settled in Western Europe, especially in Germany.

Village life

As an increasing number of young people move to the cities in search of higher-paying jobs, the populations of the mountain villages and the islands consist largely of the older people left behind. Some of those who have remained in the rural areas still live the traditional life of the Greek peasant, but others have turned away from many of the age-old customs.

In the past, Greek villagers maintained strong family ties, and each family member had a specific role and certain responsibilities. Parents made all the major decisions for their children, including selecting or approving the person their child would marry. Daughters had to marry in order of age—with the eldest wed first—and sons were allowed to marry only after all their sisters had married.

A member of the congregation kisses the hand of a Greek Orthodox priest at a religious festival. The Greek Orthodox Church is a self-governing member of the Eastern Orthodox Churches, a federation of Christian churches in Greece, the former Soviet republics, Eastern Europe, and western Asia.

A woman on the island of Crete greets her neighbor on the way back from the market.

A shopkeeper on the Ionian island of Corcyra (Kerkira or Corfu) checks his display of fresh vegetables. Many of the foods on display are ingredients of horiátiki, a traditional Greek salad.

People relax at an outdoor cafe in Iraklion, Crete's capital, largest city, chief port, and commercial center. Crete is the largest of the Greek islands.

These rigid expectations applied to husbands and wives as well. The husband took care of the household's external affairs, while the wife was primarily concerned with taking care of the house and the family.

In rural areas today, people do not follow the old customs as closely as they once did, and in the cities the old ways have all but died out. Modern lifestyles have gradually loosened the strong family traditions of the past.

However, such traditions as the Greek Orthodox festivals remain an important part of Greek life. Almost every city, town, and village has a patron saint, and festivals celebrate the saint's yearly feast day. The people enjoy food and wine after a church service, and there is singing and dancing far into the night.

Some people dress in colorful national costumes during the festivals. The men wear heavily braided jackets and pleated kilts over woolen tights, and the women wear long, brightly colored skirts and full-sleeved white blouses.

Greek food

Eating and drinking has always been an important part of Greek hospitality. Even in poor village homes, guests are welcomed with something sweet to eat, a cup of strong coffee, and a glass of cold water.

Greeks generally eat their main meal in the middle of the day. It may include such popular Greek dishes as *soupa avgolemono* (lemon-flavored chicken soup), *dolmathes* (vine leaves filled with rice and ground meat), and *souvlaki* (meat cooked on a long skewer, usually with onions and tomatoes).

The Greeks eat dinner very late at night—usually between about 8 and 11. This late dinner usually is a light meal of salad, cheese, fresh fruit, and a glass of wine.

ATHENS

Athens, one of the world's most historic cities, is a unique combination of past and present. Amid its modern factories, high-rise apartments, shopping centers, and restaurants stand the reminders of the city's ancient beginnings, when Athens was the cradle of Western civilization.

The voice of the great philosopher Socrates once rang through the Agora, the ancient hub of Athens' public life, where he taught philosophy some 2,400 years ago. In the open-air Odeon, where Western European comedy and drama were first developed by ancient Greek playwrights, present-day actors play their parts. And high above the city, on a rocky hill called the Acropolis, stands the majestic Parthenon—an ancient temple dedicated to Athena, the city's patron goddess.

The presence of these ancient ruins gives Athens a truly timeless spirit. They are reminders of the Golden Age of Greece and its magnificent achievements in science, government, philosophy, and the arts—achievements that still influence our lives today.

The ancient city

Historians know little about the history of Athens before about 1900 B.C., when the Greeks first occupied Attica, a peninsula that extends from southeastern Greece into the Aegean Sea. On and around a great flat-topped hill covering slightly more than 10 acres (4 hectares), the Greeks built a city. The hill became known as the Acropolis, from the Greek words *akro* (high) and *polis* (city), and the city became known as Athens, for its patron goddess.

In 480 B.C., most of the buildings on the Acropolis were destroyed by an invading Persian army. But by 447 B.C., the Athenians began to rebuild under the leadership of Pericles. The structures of this period, which is known as the Golden Age of Greece, still dominate the Acropolis. The greatest of these structures—and perhaps the best example of ancient Greek architecture—is the Parthenon.

The Parthenon was originally decorated with brightly painted sculptures that illustrated important events in the life of Athena, but the colors faded centuries ago.

On the Acropolis, high above the city of Athens, stand the ruins of the majestic Parthenon. Built between 447 and 432 B.C., this ancient Greek temple once held a huge gold and ivory statue of the goddess Athena, to whom the temple was dedicated.

A street vendor in Omonia Square in central Athens sells Koulouri bread rolls. The rolls resemble bagels. They are made from dough that is shaped into long ropes, then twisted into circles. They are allowed to rise again, then sprinkled with sesame seeds and baked.

Then, like other buildings on the Acropolis, the Parthenon suffered serious damage when Greece was part of the Ottoman Empire. The Ottomans used it for storing gunpowder, which exploded and destroyed the central part of the building.

In 1802, Lord Elgin, the British ambassador to Constantinople, began collecting some of the Parthenon's finest sculptures. Between 1803 and 1812, he shipped his collection to England. Today, the collection, known as the Elgin Marbles, remains on display at the British Museum, in spite of Greek demands for its return.

The city then and now

Although the Acropolis is the most striking reminder of Athens's glorious past, historic sites can be found throughout the city. At the famous old markets of Monastiraki Square, blacksmiths work their trade just as their ancestors did 2,600 years ago.

Around the corner stand the ruins of the Agora and many of the Agora's buildings, including its *stoas* (covered arcades). This marketplace was once the center for trade, athletic displays, dramatic competitions, and philosophical discussions.

Modern Athens—the leading cultural, economic, and financial center of Greece—has grown up around its historic treasures. The city's National Archaeological Museum houses masterpieces of ancient Greek jewelry, pottery, and sculpture. Its factories manufacture cement, chemicals, clothing, and other products. And its thriving tourist industry welcomes visitors from all over the world.

There is much in modern Athenian life to delight the visitor—from a high-spirited conversation in the traditional *kafenions* (coffee houses) to a sampling of delicious pastries in the *tavernas* (cafes). The city's activities center around its three main squares—Syntagma, Omonoia, and Monastiraki. But no matter where one travels in Athens, the past is always present and alive. Each monument symbolizes the rich intellectual and artistic spirit that gave birth to this city and lives on in the hearts of its people today.

Athens, the largest city in Greece, is located about 5 miles (8 kilometers) from the seaport of Piraeus. Athens was home to many of the world's great writers, philosophers, and artists. Some of the masterpieces they left behind can still be seen, including the Parthenon, the Erechtheum, and the Temple of Olympian Zeus. Many other ancient treasures are displayed in the city's museums.

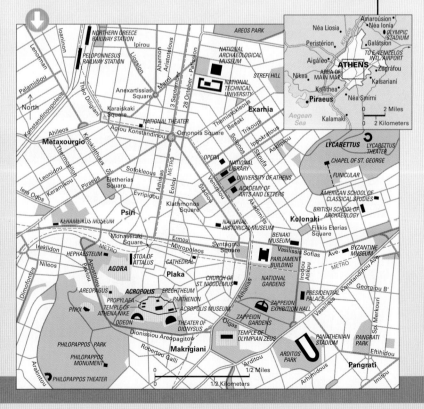

ECONOMY

The economy of Greece was almost destroyed during World War II (1939-1945) and during the Greek civil war (1946-1949). The economy expanded greatly from the 1950's until a government deficit threatened it in the first decade of the 2000's. Despite emergency loans from the European Union and the International Monetary Fund in 2010 and 2011, the crisis continued and the standard of living for most Greeks declined.

Service industries account for more than 70 percent of Greece's gross domestic product (GDP). Such industries employ most of the country's workers. The leading service industry groups include finance, insurance, and business services, as well as trade, restaurants, and hotels. Community, government, and personal services are also important to Greece's economy. Tourism benefits many of Greece's service industries.

Agriculture

Because of the mountainous landscape and the scarcity of fertile soil, only about 30 percent of Greece's entire land area is suitable for farming. Corn and wheat are Greece's main field crops. The nation's farmers also raise barley, cotton, potatoes, and tomatoes. Such fruits as grapes, oranges, and watermelons are also important. Greece is among the world's leading producers of olives, peaches, and raisins.

For many generations, family-owned land in Greece was divided among the members of each succeeding generation. As a result, Greek farms generally average less than 12 acres (5 hectares) in size. Not only does the size of the land limit production, but most farmers use old-fashioned methods and tools because the hilly terrain makes it difficult to use modern equipment.

About 35 percent of Greece's land consists of pastures and meadows, where cattle, sheep, and goats are grazed. The quality of Greece's livestock is generally poor. However, Greeks have imported higher-quality breeds from other countries to improve the stock.

The Greek fishing industry has suffered from a drastic reduction in the catch off coastal waters. The industry has seriously declined, and Greece must now import large quantities of fish to meet domestic demands.

A shopkeeper takes stock of his selection of cheeses. Because sheep and goats outnumber cattle on Greek pastures, most Greek cheeses are made from the milk of goats or sheep. Feta is the most popular cheese.

Grapes are spread on brown paper to dry in this vineyard on Crete. In about 10 to 14 days, the grapes will become raisins. The same hot sun that parches the earth provides ideal conditions for cultivating raisins.

The whitewashed walls of a country church stand out against the green trees in a Greek olive grove. The olive is one of the few crops that thrive in the thin, stony soil of Greece.

Large ships look like toy boats against the towering walls of the Corinth Canal. This deep channel—just wide enough for a single ship—crosses the land that connects the Peloponnesus with the northern mainland.

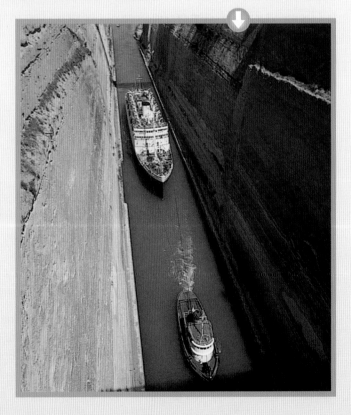

Industry

Industrial development in Greece is limited by the country's lack of raw materials. Mineral deposits are varied but limited. The nation's major mining product is a low-quality brown coal called *lignite*, which is used to generate electricity. Other important minerals include barite, bauxite, chromite, clays, iron ore and pyrite, lead, magnesite, marble, nickel, and perlite.

In order to develop its manufacturing industry, Greece needs much more electric power than it now has. The country has no commercially important deposits of natural gas, and its only significant petroleum deposit is located in the northern Aegean Sea, where geological characteristics make it difficult to drill for oil.

Today, Greece's manufacturing industry consists primarily of cement, chemicals, cigarettes, fertilizers, footwear, processed foods and beverages, and textiles. The main centers of industrial activity are Athens, Thessaloniki, and Patrai (Patras).

HISTORY

Alexander the Great (356–323 B.C.)

The first important civilization in the region that is now Greece arose on the Aegean island of Crete about 3000 B.C. Known as the Minoan culture, it flourished until about 1450 B.C. About 2000 B.C., settlers from the north began to develop small farming villages on the mainland of Greece. The culture they developed is called Mycenaean, after the large and powerful town of Mycenae in the Peloponnesus.

The Mycenaeans were in contact with the Minoan culture on Crete and adopted some aspects of that culture, such as their system of writing. Shortly after 1200 B.C., Mycenae and most other settlements in the Peloponnesus were destroyed, although no one knows for sure why this happened.

Soon, a people called the Dorians moved into the region from northern Greece, and many of the Mycenaeans fled to Asia Minor. The Peloponnesus entered a period known as the Dark Age, where people lived in isolated settlements and lost the knowledge of writing.

It was during the Dark Age that independent city-states began to develop. At first, the city-states were ruled by kings, but by about 750 B.C., the nobles in most city-states had overthrown the kings and seized power.

Beginning in 477 B.C., the city-state of Athens reached the height of its power and prosperity as the center of culture in the Greek world. The Golden Age ended when Athens and Sparta, its rival city-state, went to war in 431 B.C. The prolonged struggle, known as the Peloponnesian War, lasted until 404 B.C., ending with the surrender of the Athenians.

The Hellenistic Age

Continuing warfare between the city-states caused them to become so weak that in 338 B.C., they lost their independence to invading armies from Macedonia, a country to the north of Greece. The Macedonian ruler Alexander the Great spread the Greek culture throughout his vast empire, from Greece to India.

After Alexander's death in 323 B.C., Greece entered a period known as the Hellenistic Age, when Greek ideas continued to spread throughout Alexander's empire. When the Romans took control of the city-states in the 140's B.C., they adopted much of the Greek way of life and spread it throughout their empire.

c. 3000 B.C.	Minoan culture develops on Crete.
1600-1200 B.C.	Mycenaean culture develops on Greek mainland.
776 B.C.	The first recorded Olympic Games take place.
700's-500's B.C.	Greek colonists establish city-states.
490 and 479 B.C.	Greeks defeat Persian invaders.
461-429 B.C.	Pericles rules as leading Athenian statesman in the Golden Age of Greece.
431-404 B.C.	Athens and Sparta fight each other in the Peloponnesian War; Sparta defeats Athens.
338 B.C.	Philip II of Macedonia conquers the Greeks.
334–326 B.C.	Alexander the Great, ruler of Greece and Macedonia, conquers the Persian Empire.
323 B.C.	Greece's Hellenistic Age begins.
146 B.C.	Greece is conquered by the Romans.
A.D. 395	Greece becomes part of the East Roman (Byzantine) Empire.
1453	The Ottomans conquer Constantinople.
1821-1829	The Ottomans are defeated in the Greek War of Independence, and Greece is formed.
1833	Otto, a Bavarian prince, becomes the first king of modern Greece.
1844	Greece becomes a constitutional monarchy.
1909-1910	Military revolt leads to major reforms.
1912-1913	Greece gains territory in the Balkan Wars.
1917-1918	Greece fights in World War I on the side of the Allies.
1922	Ottomans defeat Greek forces in Asia Minor.
1924	Greece is declared a republic.
1935	Constitutional monarchy is restored in Greece.
1936-1941	General Joannes Metaxas rules as dictator.
1941-1944	Axis forces occupy Greece during World War II.
1952	Greece joins NATO.
1960	Cyprus becomes independent.
1967	Army officers seize the Greek government.
1973	Premier George Papadopoulos abolishes monarchy. Later, military officers overthrow Papadopoulos government.
1974	Greece holds free elections, and a civilian government is formed.
1993	Greece and other members of the European Community form the European Union.
2009	George Andreas Papandreou follows in his father's and grandfather's footsteps and becomes prime minister of Greece.

The famous Lion Gate leads to the Acropolis of Mycenae, the first major civilization to develop on the Greek mainland. A massive beam across the top supports a triangular bas-relief of two rearing lionesses.

Plato
(427?-347? B.C.)

George Papandreou
(1952-   )

The Tholos (rotunda) at Delphi was built shortly after 400 B.C. Delphi, the oldest and most influential religious site in ancient Greece, was home to the famous oracle (prophet). The Greeks believed that the oracle spoke the words of Apollo.

When the Roman Empire was divided in A.D. 395, Greece became part of the East Roman Empire—also known as the Byzantine Empire. Despite centuries of war and invasions from neighboring peoples, the Byzantine Empire continued to control at least part of Greece for over 1,000 years. The capture of the Byzantine capital of Constantinople by the Ottomans in 1453 marked the end of the Byzantine Empire and of Greek independence for nearly four centuries.

The making of modern Greece

Although a Greek independence movement developed in the 1700's, the Greek War of Independence did not begin until 1821. Six years later, the United Kingdom, France, and Russia sent troops to end the fighting and establish Greece as a self-governing country. In 1833, a Bavarian prince, Otto, became the first king of Greece.

The new kingdom was less than half the size of present-day Greece. About 3 million Greeks lived in Ottoman territory, and 200,000 lived in the British-controlled Ionian islands. The resulting economic confusion and political discontent led to a peaceful revolution in 1862 that forced Otto I to give up his throne. He was replaced by a Danish prince who became George I.

In 1912–1913, Greece and several other Balkan states defeated the Ottomans in the First Balkan War. The Balkan states then fought against each other during the Second Balkan War, in 1913. As a result of these wars, Greece gained the island of Crete, southern Epirus, part of Macedonia, and many Aegean islands.

In 1924, Greece declared itself a republic, which lasted until 1935, when the monarchy was temporarily restored. A new Greek constitution in 1975 made the nation a parliamentary republic.

ANCIENT GREEK CIVILIZATION

The modern world owes much to the ancient Greeks. The ancient Greeks developed the first democratic government; drama was born in their huge, open-air theaters; and Greek thinkers developed the reasoning skills necessary for demonstrating important mathematical principles. The splendid Olympic Games—the world's most honored sporting tradition—first took place in the Stadium of Olympia in Greece.

The magnificent achievements of the ancient Greeks are still with us today—in the majestic temples of the Acropolis, in the poetry of Homer's *Iliad* and *Odyssey,* and in our modern Olympic games. But perhaps most important, the ancient Greeks encouraged creative thinking, valued personal freedom, and explored the human potential. They laid the foundation for the continuing search for knowledge, truth, and new forms of expression throughout generations of Western civilization.

The ruins of the Temple of Poseidon on Cape Sounion overlook the island of Salamis, near the site of a great sea battle between the Greeks and the Persians in 480 B.C. That Greek victory helped save the country from being invaded by the Persians.

The Acropolis of Athens was a rocky, flat-topped hill. Its buildings, the crowning glory of ancient Greece, date from the mid-400's B.C. Paths lead past the Temple of Athena Nike (1) and through the Propylaea (2) to the magnificent Parthenon (3). The Erechtheum (4) stands to the north, while on the south slope stand the Odeon of Herodes Atticus (5) and the Theater of Dionysus (8). Linking the two theaters is the Stoa of Eumenes II (6), where the great philosophers often walked with their students. Behind it stands another stoa, the Asklepion (7), which was dedicated to the Greek god of healing.

Aristotle was a philosopher, teacher, and scientist. He was once Plato's pupil at the Academy, a school of philosophy in Athens.

Greek civilization arose along the shores of the Aegean and Ionian seas. Ancient Greece consisted chiefly of a peninsula that separated the two seas, nearby islands, and the coast of Asia Minor (now part of Turkey).

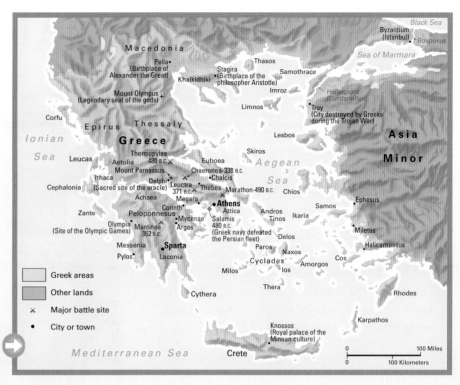

The Golden Age of Greece

During the 500's B.C., the Persian Empire conquered the Greek city-states of Asia Minor. From 499 to 494 B.C., the city of Athens aided the captured city-states in rebelling against Persian rule. Athens also played a leading role in preventing the Persians from gaining more Greek territory.

After the wars with Persia, Athens became head of the Delian League, an organization of city-states formed as a continuing defense against the Persians. The league quickly developed into the Athenian empire, and Athens became the literary and artistic center of Greece. Under the leadership of the great statesman Pericles, the Greeks enjoyed a period of outstanding cultural achievement, known as the Golden Age.

Of all the city-states in ancient Greece, Athens had the most successful democracy. Along with its advanced political system, Athens enjoyed great prosperity as an international trading center. Athenian merchant ships carried olive oil, painted pottery, wine, wool, and other goods to ports in Egypt, Sicily, and Scythia—a country on the Black Sea. There, Greek merchants would sell their goods for slaves and for such products as grain, timber, and metals.

Philosophical and artistic triumphs

Philosophy emerged in ancient Greece, and Athens nourished its most important teachers and philosophers—Socrates, Plato, and Aristotle. They often gathered with their pupils in the Agora, the marketplace in the center of the city, to discuss philosophical issues. The early philosophers wondered about the substance of the universe and how it operated. Later philosophers explored the nature of knowledge and reality.

Socrates was a teacher in Athens during the Golden Age. He believed in the basic goodness of people, and he said that evil and wrong actions arise from ignorance. Socrates taught his students by questioning them and exposing the weaknesses of their ideas. But some of the more influential citizens mistrusted his ideas, and he was sentenced to death in 399 B.C.

Plato, a friend of Socrates and one of his most gifted pupils, explored such subjects as beauty, justice, and good government. In 387 B.C., Plato founded a school of philosophy and science called the Academy. Some scholars consider the Academy the world's first university.

Aristotle, one of the greatest and most influential thinkers in Western culture, was a student at the Academy. In his writings, Aristotle summed up the rich intellectual tradition he had inherited from his teachers.

The ancient Greeks excelled in the arts as well as in philosophy. Athenian playwrights wrote comedies and tragedies that were performed at religious festivals. Greek writers introduced many new forms to the world of literature, including lyric and epic poetry, philosophical essays and dialogues, and critical and biographical history. Their writings were the model for much of the later literature in the West.

THE AEGEAN ISLANDS

Scattered in the Aegean and Ionian seas off the Greek mainland lie hundreds of small islands that make up about a fifth of Greece's land area. Many of these islands played an important role in the history and development of ancient civilization, while others are tiny islets virtually untouched by humanity.

Most of the Greek islands, which are known as the Grecian Archipelago, lie in the Aegean Sea, an arm of the Mediterranean Sea between Greece, Turkey, and the island of Crete. Some of these islands are ancient volcanoes and are made of lava, while others are made of pure white marble.

Evvoia, the largest of the central Greek islands, is located just off the coast of mainland Greece. The rest of the Aegean Islands form two main groups, the Cyclades and the Sporades. The Dodecanese Islands are part of the Sporades group.

The Cyclades

The Cyclades lie in the southern Aegean. They are so named because, to the ancient Greeks, the islands appeared to lie in a circle (*kyklos* in Greek) around the island of Dhilos, the birthplace of Apollo. The Cyclades include such well-known islands as Ios, Mikonos, Milos, Naxos, Paros, and Thira (Santorin). Geologically, the Cyclades are a continuation of the hills and mountains of the Greek mainland.

The Cyclades have been populated for more than 4,500 years. On the island of Naxos, anthropologists found evidence of an early Cycladic culture dating from about 3000 B.C. The islands were later influenced by the Minoan and Mycenaean civilizations. Their location on shipping routes between Greece and Asia Minor made the Cyclades important to the ancient Greek world.

The Sporades

The northern Sporades lie in the northeastern Aegean Sea off the coast of Asia Minor (Turkey). The larger islands of the Sporades include Khios

A group of women enjoy a chat while embroidering and crocheting on a sunny street on the island of Limnos. Today, the economy of the Aegean Islands depends largely on tourism.

(Chios), Lesbos, Limnos, and Samothraki (Samothrace). South of the Sporades, the Dodecanese include the larger islands of Rhodes and Samos, and the smaller islands of Kalimnos, Karpathos, Kos, Patmos, and Simi.

The Sporades have been Greek in culture since the 400's B.C., when Athens gained control of the islands. The islanders are proud of their role in the rise of ancient Greece. Many famous Greek thinkers, including the philosopher and mathematician Pythagoras of Samos, came from these islands. The island of Lesbos was a major cultural center from about 600 B.C. to the end of the Golden Age of Greece in 431 B.C.

Lesbos, Khios (Chios), and Samos, which lie only a short distance from the mainland of Turkey, have felt the constant presence of the Turks. The Dodecanese Islands were also influenced by Italian, German, and British rule.

Visiting the islands

The beauty and charm of the Greek islands attract many tourists. Standing out against the deep blue Aegean Sea, the white houses of the islands reflect the sun's dazzling rays. The narrow, winding streets are lined with colorful village houses—most with blue doors and many decorated with tiles. The sunny, Mediterranean climate makes walking a pleasure, while the beaches provide a refuge from the hectic routines of city life. The many ancient temples and historic monuments of the Aegean Islands also attract visitors from around the world.

Tourism makes up a major part of the economy of the Aegean Islands. The island of Rhodes in particular has developed into a major tourist center. During the tourist season, the islands are crowded with tourists and with temporary workers employed in the hotel and resort industry. But at the end of the season, the pace of island life slows considerably. Jobs are scarce, and many young people must leave the islands to seek higher-paying jobs in the mainland cities.

Gleaming white houses line the coast of the Cyclades island of Mikonos, a major tourist center. The whitewash on the houses reflects the sun's rays and keeps the interiors cool.

Lions carved from Naxian marble have guarded the Sacred Lake of Apollo on Dhilos (Delos) since the 600's B.C. Dhilos was important in ancient times because it was thought to be the birthplace of Apollo and Artemis.

The Aegean Islands, which lie in the Aegean Sea between the Greek mainland and Turkey, are rocky and sparsely populated. However, Rhodes, Delos, Tinos, Paros, Mikonos, and Siros are major tourist attractions. Another attraction is the island of Thira, which some historians believe is the lost continent of Atlantis.

CRETE AND RHODES

Crete, the largest of the Greek islands, lies about 60 miles (97 kilometers) south of the Peloponnesus. Crete is famous for its scenic landscape and colorful traditions, as well as for the Minoan ruins that still dot its landscape.

Early history

Crete has an important place in Greek history. The first major European civilization, the Minoan culture, developed on this island. The Minoans are named for Minos, the king of Crete in Greek mythology.

According to legend, Minos kept a Minotaur—a monster with the head of a bull and the body of a man—in the Labyrinth, a building designed as a maze from which no one could escape. After Minos conquered much of Greece, including Athens, he sacrificed seven Athenian youths and seven Athenian maidens to the Minotaur every year. Finally, Theseus—one of the intended victims—killed the Minotaur and eloped with Minos's daughter Ariadne.

The Minoan culture flourished on Crete from about 3000 B.C. until about 1450 B.C. Some scholars believe that the effects of a volcanic eruption on the nearby island of Thira (Santorin) may have weakened the Minoan culture. Some towns on the island were never reoccupied. The culture began to decline in the early 1300's B.C., and by the mid-1100's B.C. it had completely disappeared.

In 68 B.C., the Romans invaded Crete, and in 66 B.C. it became a Roman province. After the Roman Empire was divided in A.D. 395, Crete came under the rule of the East Roman (Byzantine) Empire. Between 1204 and 1669, Venice ruled Crete.

The Ottomans occupied Crete from 1669 to 1898 and outlawed the Christian Orthodox religion of the islanders. Many of the people fled to the mountains to take up the struggle against Turkish occupation. After the Turks were forced to leave the island, Crete was independent until it became a part of Greece in 1913.

The massive walls of the Residence of the Grand Master rise high above the town of Rhodes. The palace was built by Helion de Villeneuve in the 1300's. It was later used as a prison by the Ottomans.

Relics of ancient times

The Palace of Minos at Knossos has the most important remains of Crete's Minoan culture. The palace stands about 3 miles (5 kilometers) southeast of Iraklion, Crete's largest city. Sir Arthur Evans, a British archaeologist, began unearthing the enormous palace in 1900 and had it partially rebuilt.

Evans's discoveries showed that the king's residence was surrounded by impressive villas decorated with wallpaintings and plaster reliefs. These artworks include scenes of *bull-leaping*—young men and women leaping over the backs of bulls. Seafaring themes are also depicted.

In addition to the Minoans, the Romans, Venetians, and Muslim Turks also left their mark on the island. Rethimnon, Crete's third-largest city, is an attractive mixture of these different cultures. The old part of the town contains a Venetian castle as well as Turkish houses with their latticed wooden balconies. And on the site of Gortys, the ancient Roman capital of Crete, stand the foundations of a Roman palace.

A simple island life

The island's largely mountainous landscape makes it difficult for farmers to use modern equipment, so much of the work is done by hand. Most crops are grown in the fertile river valleys in the upland areas. Farmers in Crete grow grapes, olives, oranges, vegetables, and nuts. Some villagers make a living selling hand-crafted items, such as baskets, metalwork, and pottery.

The northern cities have a growing food-processing industry, and factories in Iraklion manufacture soap and leather goods. The tourist trade provides employment for a large number of islanders.

The people of Crete are very proud of their ancient heritage. Most of them speak Greek, belong to the Eastern Orthodox Church of Crete, and follow many age-old customs.

The setting sun casts a warm glow over the harbor at Khania (Canea). The color-ful old-town waterfront features Venetian buildings with red-tiled roofs.

Playful dolphins decorate the walls of the Queen's Apartments in the Palace of Minos at Knossos. These frescoes were painted on wet plaster, a technique that gives the colors a vibrant quality.

RHODES

Rhodes, the largest of the Dode-canese Islands, lies 12 miles (19 kilo-meters) off the southwestern coast of Turkey. Rhodes features beautiful scenery and historic buildings.

The fertile soil on Rhodes yields abundant crops of olives, tobacco, grapes, and other fruit. However, most of the people on Rhodes make their living in the tourist industry.

Rhodes was once a wealthy state of Greece, home to poets, artists, and philosophers. The Colossus of Rhodes, a huge bronze statue of the god Helios, stood in its harbor. One of the Seven Wonders of the Ancient World, it was destroyed by an earthquake in 224 B.C.

From 1310 till 1522, the island was occupied by the Knights Hospitallers of St. John, crusaders who sought to remove the Muslims from Jerusalem. Turkish forces held Rhodes from 1522 till 1911, when Italy took it. Rhodes became part of Greece in 1947.

MOUNT ATHOS

According to legend, a ship carrying a holy man set sail along the northeast coast of what is now Greece. Just as the vessel passed by the northeastern peninsula where Mount Athos stood, the ship became stuck in the water and would not budge.

Suddenly, the holy man, whose name was Peter, declared that God had called him to Mount Athos. He jumped off the ship at present-day Karavostasi (Bay of Standing Ships) and swam ashore. As the astonished sailors watched Peter climb the mountain, their ship once again sailed free.

It is said that Saint Peter of Mount Athos lived in a tiny cave on Mount Athos for 50 years during the A.D. 700's. As time passed, other Christian hermits also came to live in the area. The first monastery of Athos was probably founded in 963.

Over the next 200 years, 40 monasteries were established, and Athos became a center for Christian Orthodox learning. Monks from Russia, Romania, and Bulgaria came to study and live in Athos.

Today, Athos is a self-governing monastic republic, where monks live in 20 monasteries. It lies on the easternmost prong of the Khalkidhiki Peninsula. Set amid the breathtaking beauty of the untouched Mediterranean coast, the region covers about 130 square miles (335 square kilometers). Mount Athos, known to the Greeks as *Ayion Oros* (Holy Mountain) dominates the rocky peninsula, rising to a height of 6,670 feet (2,033 meters).

Athos is governed by the Holy Community, located in Karyas. The Holy Community is a group of 20 representatives who are elected annually by the monastic communities. A governor appointed by the Greek government is responsible for civil administration on Mount Athos.

The monks' republic

Most of the monasteries on Athos were founded in the 900's and the 1000's. Clinging to the edges of steep, inaccessible cliffs, the monasteries were easy to defend against pirates. Today, their well-preserved Byzantine and medieval architecture, framed by lush vegetation and dense woodlands, attracts scholars and tourists alike.

The libraries of Athos hold many elaborate Byzantine manuscripts and sacred objects. Services take place in cross-shaped chapels decorated by frescoes and mosaics and lined with religious statues set with silver and precious stones.

A serious problem confronting Athos today is the decreasing number of new monks. The monastic population, once as high as 40,000, has declined to 2,000, close to the minimum required to keep the monasteries functioning. The upkeep and repair of the old stone buildings require a great deal of time. In addition, the lack of manpower has been blamed for fires that have destroyed entire wings of monasteries.

With the occasional help of Greek farmworkers, the monasteries support themselves by producing olives, vegetables, and wine. The monks also keep chickens and cats. Some of the monks also produce handicrafts and woodcarvings, which are sold to visitors. The monasteries also receive income from lands in Greece and from private donations.

The domestic tasks of the monks are as much a part of their routine as their hourly prayers. The monks hope to free themselves from worldly concerns and become closer to God through worship and labor.

The Monastery of St. Gregorius clings to a cliff on the slopes of Mount Athos. Many Athos monks live outside the monasteries in sketes (semi-independent communities) scattered throughout the Khalkidhiki Peninsula.

Christian hermits probably have lived on Athos since the 700's, when legend says that Saint Peter of Mount Athos arrived. Some monks still live as hermits in caves scattered throughout the forest.

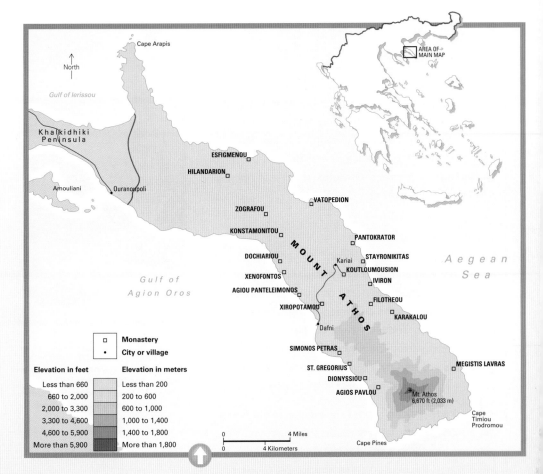

North

Cape Arapis

Gulf of Ierissou

Khalkidhiki Peninsula

Amouliani

Ouranoupoli

ESFIGMENOU

HILANDARION

VATOPEDION

ZOGRAFOU

KONSTAMONITOU

PANTOKRATOR

DOCHIARIOU

Kariai

STAYRONIKITAS

KOUTLOUMOUSION

XENOFONTOS

IVIRON

AGIOU PANTELEIMONOS

FILOTHEOU

XIROPOTAMOU

KARAKALOU

Dafni

SIMONOS PETRAS

ST. GREGORIUS

DIONYSSIOU

MEGISTIS LAVRAS

AGIOS PAVLOU

Mt. Athos
6,670 ft (2,033 m)

Cape Timiou Prodromou

Cape Pines

Gulf of Agion Oros

M O U N T   A T H O S

A e g e a n   S e a

AREA OF MAIN MAP

□ Monastery
• City or village

Elevation in feet	Elevation in meters
Less than 660	Less than 200
660 to 2,000	200 to 600
2,000 to 3,300	600 to 1,000
3,300 to 4,600	1,000 to 1,400
4,600 to 5,900	1,400 to 1,800
More than 5,900	More than 1,800

0   4 Miles
0   4 Kilometers

Athos lies on the easternmost prong of the Khalkidhiki Peninsula on the Mediterranean coast. The region covers about 130 square miles (335 square kilometers). Mount Athos, dominates the rocky peninsula, rising 6,670 feet (2,033 meters).

Visiting Athos

Women and children are not permitted to set foot on Mount Athos beyond the town of Ouranoupoli. Male visitors who have obtained permission from the authorities in Athens or Thessaloniki (Salonika) usually take a boat to Dhani. Once on Athos, most travel is done on foot or by mule. Upon arrival, visitors are reminded to follow the rules and behave in a respectful manner—smoking, singing, and whistling are frowned upon, for example.

To outsiders, life on Athos is a world away from their own experience. Most of the monasteries follow the Julian calendar, which is 13 days behind the calendar used by the rest of the world. And most still use an old Byzantine method of reckoning time, in which their days begin at sunset, when clocks read 12 o'clock. At that time, the gates close and everyone retires to their quarters. The drumbeat that calls the monks to prayer can be heard throughout the night.

GREENLAND

The largest island in the world, Greenland is situated in the North Atlantic Ocean, only about 10 miles (16 kilometers) from Canada. Although it is geographically part of North America, Greenland is a province of Denmark, with the constitutional right of *home rule* (local self-government).

Landscape of ice

Viking explorers gave Greenland its name, hoping to attract Norwegian settlers to the island. However, most of Greenland lies north of the Arctic Circle, and only the coastal areas are green during the island's short summers. The landscape consists of a low inland plateau surrounded by coastal mountains. The plateau is covered by a thick, permanent icecap averaging 1 mile (1.6 kilometers) in thickness.

Hundreds of long, narrow sea inlets called *fiords* penetrate the coastal mountains that surround the plateau, providing a transportation link between the interior and the coastal cities and towns. Glaciers, formed on the icecap by compressed snow, flow slowly down coastal valleys and into the fiords, often breaking up into enormous icebergs.

Settlement and development

The first Greenlanders were Arctic tribes of the Sarqaq culture, who came to the island from North America. Norwegian Vikings are believed to have sighted what is now Greenland about A.D. 875. The Viking explorer Erik the Red brought the first settlers to the island about 985.

In 1261, the Greenland colonists voted to join Norway, and when Norway united with Denmark in 1380, the island came under Danish rule. The colony died out during the 1400's, and Greenland was lost to the outside world for many years. It was rediscovered in 1578 by the English navigator Martin Frobisher, whose expedition had set out to find the Northwest Passage. The colonization of the island began again in 1721, when Hans Egede, a Norwegian missionary, established a mission and trading center near what is now the capital city of Nuuk.

Snow blankets wooden houses in Upernavik, an island off the southwestern coast of Greenland. Almost all Greenlanders live along the country's southwestern coast.

The busy seaport of Umanak is one of Greenland's largest towns. Like most of the other settlements, Umanak lies on the western coast, the warmest region of Greenland.

When the union between Norway and Denmark ended in 1814, Greenland remained with Denmark. Norway disputed the move, but in 1933, a world court upheld Denmark's claim. In 1953, a new Danish constitution changed Greenland from a colony to a province.

In 2009, Greenland gained complete responsibility for its internal affairs. The Danish government, however, remains responsible for Greenland's foreign affairs and defense.

A changing environment

The climate of Greenland is very cold, but it has been gradually getting warmer since the early 1900's. The warming of the sea has brought great numbers of fish to Greenland's coastal waters, while causing seal herds to migrate northward. This shift in natural patterns led the Danish government to promote a change in Greenland's economic base from seal hunting to fishing. More than 30 percent of Greenland's people now work in the fishing industry.

The tundra landscape of Hare Fjord stretches into the distant mountains. The fjord is part of the Scoresby Sund fjord system and is home to many varieties of wildlife throughout the year.

Greenland's location has great strategic importance. Scientists on the island can forecast storms on the North Atlantic Ocean, and U.S. military bases there form a major part of the North American defense system.

More than 80 percent of the people of Greenland were born there, and most of the others are Danish immigrants. Relatively few Greenlanders have entirely Inuit (Eskimo) ancestry. Most have Danish ancestors as well and no longer follow the old Inuit ways. They now live mainly in towns, where they work in the fishing industry, live in wooden houses, and wear European clothing.

Most of the Greenlanders who have mainly Inuit ancestry live in the far northwest regions and hunt seals. Most of these people follow the traditional Inuit way of life. They use the meat of the seals for food, the seals' blubber for oil, and the animals' skins to make clothing and kayaks.

GRENADA

Grenada is the most southerly of the Windward Islands, which are part of the Lesser Antilles. It lies in the Caribbean Sea about 90 miles (140 kilometers) north of Venezuela. The nation of Grenada includes the island of Grenada, which makes up most of the country, and several tiny islands nearby. It also includes Carriacou, which lies about 17 miles (27 kilometers) northeast of the main island.

Island of Spices

Grenada is a mountainous volcanic island, with thickly forested land and many gorges and waterfalls. Its highest point is Mount Saint Catherine, an extinct volcano that towers 2,756 feet (840 meters) above the countryside.

Along the coast, Grenada's magnificent beaches include the famous Grand Anse Beach, which stretches about 2 miles (3 kilometers) along the southwest part of the island. In addition to its scenic landscape, Grenada's pleasant climate attracts many tourists to this small island paradise.

Grenada is a developing country, and its economy is based on agriculture and tourism. As a leading producer of spices—especially nutmeg—Grenada is often called the *Island of Spices*. Other crops include bananas and cocoa.

About 95 percent of Grenada's people have African or mixed African and European ancestry. The official language is English, but many people speak an English or French dialect.

History

In 1498, when Christopher Columbus landed in what is now Grenada, he found Carib Indians living there. Columbus named the island *Concepcion*, but later explorers called it Grenada. In 1650, the French claimed Grenada and later slaughtered many Indians.

France and Great Britain fought for control of Grenada until the island became a British colony in 1783. In the mid-1900's, the British gave Grenada some control over its own affairs. In the early 1970's, Prime Minister Eric M. Gairy led a movement for independence from the United Kingdom.

FACTS

Official name:	Grenada
Capital:	St. George's
Terrain:	Volcanic in origin with central mountains
Area:	133 mi² (344 km²)
Climate:	Tropical; tempered by northeast trade winds
Main river:	Great
Highest elevation:	Mount Saint Catherine, 2,756 ft (840 m)
Lowest elevation:	Caribbean Sea, sea level
Form of government:	Constitutional monarchy
Head of state:	British monarch, represented by governor general
Head of government:	Prime minister
Administrative areas:	6 parishes, 1 dependency
Legislature:	Parliament consisting of the Senate with 13 members and the House of Representatives with 15 members serving five-year terms
Court system:	Eastern Caribbean Supreme Court
Armed forces:	N/A
National holiday:	Independence Day - February 7 (1974)
Estimated 2010 population:	110,000
Population density:	827 persons per mi² (320 per km²)
Population distribution:	69% rural, 31% urban
Life expectancy in years:	Male, 65; female, 68
Doctors per 1,000 people:	1.0
Birth rate per 1,000:	20
Death rate per 1,000:	6
Infant mortality:	15 deaths per 1,000 live births
Age structure:	0-14: 30%; 15-64: 65%; 65 and over: 5%
Internet users per 100 people:	22
Internet code:	.gd
Languages spoken:	English (official), French patois
Religions:	Roman Catholic 53%, Anglican 13.8%, other (primarily Protestant) 33.2%
Currency:	East Caribbean dollar
Gross domestic product (GDP) in 2008:	$639 million U.S.
Real annual growth rate (2008):	3.7%
GDP per capita (2008):	$6,028 U.S.
Goods exported:	Bananas, cocoa, fish, nutmeg, transportation equipment
Goods imported:	Chemicals, food, machinery, motor vehicles, petroleum products
Trading partners:	Canada, Caribbean countries, United Kingdom, United States

Grenada gained independence in 1974, and the new country became a constitutional monarchy. Gairy served as prime minister until 1979, when rebels led by Maurice Bishop overthrew his government. Bishop was a Marxist who had close ties to Cuba, but some rebels felt that Bishop did not go far enough in adopting a complete Marxist system in Grenada. In October 1983, they took over the government and killed him.

Soon after Bishop's death, a number of other Caribbean nations—Antigua and Barbuda, Barbados, Dominica, Jamaica, St. Lucia, and St. Vincent and the Grenadines—called upon the United States to help restore order in Grenada. They feared that Grenada would be used as a base by Cuba and the Soviet Union to support terrorism and revolution throughout Latin America.

On Oct. 25, 1983—two days after Bishop was killed—U.S. troops invaded Grenada. Small numbers of troops from six Caribbean nations also took part in the invasion. After several days, the multinational force took complete control of the country. By December 15, all U.S. troops had been pulled out of the country. In elections held in 1984, a centrist coalition called the New National Party (NNP) won a clear majority. Its leader, Herbert A. Blaize, became prime minister and set about restoring stability to his country. The National Democratic Congress (NDC) formed the opposition.

Before he died in 1989, Blaize formed a new party, The National Party (TNP). The NDC won the 1990 elections and formed a governing coalition. The NNP took control after the 1995 election and governed for the next 13 years. The NDC won the election in 2008.

St. George's is Grenada's capital, chief port, and commercial center. The city lies on a picturesque harbor along Grand Anse Bay on the southwest coast of the country.

The island nation of Grenada consists of the main island of Grenada and the southern islands of the Grenadines. Grenada became independent in 1974. Grenada suffered many political problems during the late 1970's and early 1980's.

Guadeloupe, a group of islands in the Lesser Antilles, consists of two main islands, a smaller island group called Îles des Saintes, and five small islands. Together, the islands make up an overseas *department* (administrative district) of France within the French Community. Guadeloupe covers 630 square miles (1,631 square kilometers).

The larger of the two main islands—Guadeloupe, or Basse-Terre—is separated from the other main island—Grande-Terre—by Rivière Salée, a narrow, bridged strait. From the air, Basse-Terre and Grande-Terre resemble a butterfly. The five small islands are Marie-Galante, La Désirade, St.-Barthélemy, the northern part of St. Martin, and Petite-Terre. The town of Basse-Terre, on Basse-Terre Island, is the capital of Guadeloupe.

The picturesque island of Terre-de-Haut attracts visitors with its beautiful beaches and bays, fine snorkeling, and historical sites. Terre-de-Haut is one of a group of small islands southwest of Guadeloupe called Les Saintes.

A worker harvests sugar cane on the island of Grande-Terre. Agriculture is the major source of income for Guadeloupe. Along with sugar cane, leading agricultural products include bananas, cocoa, and coffee.

Guadeloupe has a population of about 416,000. Most of the people are of mixed African and European ancestry. A group of descendants of the original Norman and Breton settlers lives in the Îles des Saintes group.

Agriculture provides the chief source of income in Guadeloupe. Leading farm products include bananas, cocoa, coffee, and sugar cane. Farmers also grow vegetables and tobacco for local markets. Several distilleries export rum.

Tourism provides another source of income for Guadeloupe. An international airport is located on the island of Grande-Terre.

History and government

Warlike Carib Indians were living in Guadeloupe when Christopher Columbus reached the islands in 1493. The Carib resisted European settlement until 1635, when the first French settlers arrived. Since that time, Guadeloupe has remained a French possession, except when the British occupied the territory between 1759 and 1813.

The islands of Guadeloupe lie in the Caribbean Sea about 370 miles (595 kilometers) north of Venezuela. Guadeloupe has a hot, damp climate from June to December, but the trade winds tend to moderate the heat. From January to May, the islands enjoy cool, dry weather. Dense tropical vegetation blankets much of central Basse-Terre, the mountainous western island of Guadeloupe. Along the island's west coast, divers enjoy some of the finest coral reefs in the southern Caribbean.

Islanders run for shelter from the fierce waves caused by Hurricane Hugo, which swept across Guadeloupe in 1989. Hugo left a long trail of destruction across the islands. Thousands of people lost their homes in the devastating storm.

Guadeloupe became a French overseas department in 1946. A general council of elected members governs the department, and deputies represent the group in the French National Assembly. France is also Guadeloupe's chief trading partner, representing the majority of the trade on the islands.

Grande-Terre

Part of the outer arc of the Lesser Antilles, Grande-Terre consists of a low plateau with *karst* (limestone) formations. Dense rain forests once covered the island, but now the landscape is dominated by large plantations and small villages where the plantation workers live.

A sandy beach about 25 miles (40 kilometers) long stretches along the Caribbean coast of Grand-Terre between Point-à-Pitre and the Pointe-des-Châteaux, the easternmost tip of the island. Along this coast, a number of outstanding resorts welcome sunbathers and water-sports enthusiasts. The main center of activity on the southern coast is Gosier, where vacationers enjoy fine restaurants and exciting nightlife.

In Point-à-Pitre, Guadeloupe's chief port and largest city, the streets are lined with peddlers selling exotic fruits, spices, and seafood of all kinds, along with mysterious ingredients for folk medicines. The harbor is another center of activity, as oceangoing ships arrive with goods from France and other countries. Bananas, rum, and sugar cane are then loaded back on board for the return trip.

Although Point-à-Pitre has suffered fires, earthquakes, enemy attacks, and hurricanes, many buildings dating from 1900 still survive. Their carved wooden facades and elaborate wrought-iron balconies display the charm of French colonial architecture.

Basse-Terre

Unlike Grand-Terre, Basse-Terre is a volcanic island. The Indian name for Basse-Terre is *Karukera*, which means *Isle of Beautiful Waters*. The name refers to the island's many rivers, lakes, and waterfalls. Near the west coast, divers can enjoy beautiful coral reefs and schools of tropical fish in the Caribbean. Volcanic mountains, including the active Soufrière, extend across Basse-Terre from north to south. In the central part lies a nature park, where a road leads partway up to the volcano. Tropical rain forests, exotic flowering plants, hot springs, cinder cones, and sulfur fields cover the land.

GUAM AND THE NORTHERN MARIANAS

The Mariana Islands are the northernmost islands of Micronesia. Island residents are U.S. citizens. All the islands in this group, except Guam, are a commonwealth of the United States called the Commonwealth of the Northern Marianas. The government of the commonwealth controls its internal affairs, but the United States remains responsible for the island's foreign affairs and defense.

Guam, which lies at the south end of the Marianas about 1,300 miles (2,100 kilometers) east of the Philippines, is a territory of the United States and serves as a vital air and naval base in the Pacific Ocean. English is the official language, but most people speak Chamorro, a native language. The people of Guam elect a delegate to the U.S. House of Representatives. The delegate can vote in House committees, but not on the House floor.

Land

The Marianas are the southern part of a submerged mountain range that extends 1,565 miles (2,519 kilometers) from Guam almost to Japan. The summits of 15 volcanic mountains in this range form the Marianas. The 10 northern islands have a rugged landscape, and some have volcanoes that erupt periodically. The five southern islands have limestone or reef rock terraces on volcanic slopes that show they are older than the northern group. Guam is the largest of the southern islands.

Coral reefs lie off the coast of Guam. Many forests in northern Guam have been cleared for farms and airfields. The south has a range of volcanic mountains where rivers originate. Earthquakes occasionally strike the island. Guam is warm most of the year, and annual rainfall averages 90 inches (230 centimeters). Typhoons frequently hit the island.

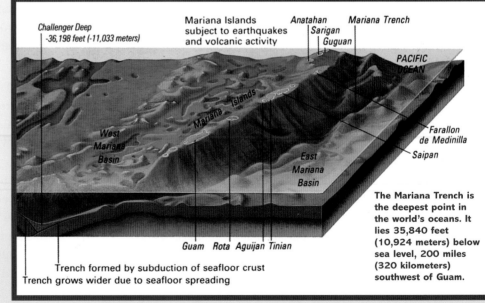

Challenger Deep
-36,198 feet (-11,033 meters)

Mariana Islands subject to earthquakes and volcanic activity

Anatahan
Sarigan
Guguan

Mariana Trench

PACIFIC OCEAN

Mariana Islands

West Mariana Basin

East Mariana Basin

Farallon de Medinilla

Saipan

Guam Rota Aguijan Tinian

Trench formed by subduction of seafloor crust
Trench grows wider due to seafloor spreading

The Mariana Trench is the deepest point in the world's oceans. It lies 35,840 feet (10,924 meters) below sea level, 200 miles (320 kilometers) southwest of Guam.

Hagåtña, the capital of Gaum, lies on the west coast of the island. Gleaming white modern buildings stand along the city's beachfront.

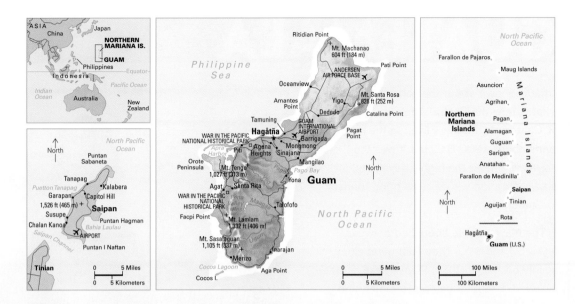

The Northern Marianas and Guam make up the Mariana Islands. They have a land area of 391 square miles (1,013 square kilometers), formed by the summits of 15 volcanic mountains.

A Micronesian girl performs a dance at a cultural festival on the island of Saipan in the Northern Marianas Islands. These islands are the northernmost islands of Micronesia.

People and history

The native islanders of the Marianas are called *Chamorros*. Their ancestors, among the earliest settlers of Micronesia, arrived from Asia thousands of years ago. The Chamorros, who have intermarried with Europeans, Filipinos, and other peoples, now practice many Western customs.

The Marianas have a population of about 271,000. About 181,000 of the people live on Guam, and about 90,000 live on Saipan, the capital and largest island of the Northern Marianas. Many of the people on Guam are U.S. military personnel and their dependents. The island's major sources of income are the U.S. military service and tourism. Tourism is an important economic activity on Saipan as well.

The Portuguese explorer and navigator Ferdinand Magellan led the first European expedition to the area in 1521. The islands received their name, however, from Spanish Jesuits who arrived in 1668. Spain governed the islands from 1668 to 1898.

After the Spanish American War (1898), the United States kept Guam as a naval base. Spain sold the rest of the Marianas to Germany. Japan seized the German-held islands at the start of World War I (1914–1918), and a League of Nations mandate gave Japan control of the Marianas after the war. The islands were the scene of heavy fighting during World War II (1939–1945).

Japan captured Guam in 1941. In 1944, U.S. forces recaptured Guam and took control of the Northern Marianas. In 1954, the Strategic Air Command of the U.S. Air Force established Andersen Air Force Base and made Guam its Pacific headquarters. In 1950, the United States declared Guam a territory. In 1986, an agreement to make the Northern Marianas a self-governing commonwealth went into effect.

GUATEMALA

Little is known about the earliest people in what is now the Central American country of Guatemala. Farmers lived in the Highlands about 1000 B.C., but it was the Mayas who built the highly developed Indian civilization whose ruins bring tourists to Guatemala.

The Maya people flourished between A.D. 250 and 900. They built great cities, mainly on the Northern Plain. These cities included beautiful palaces, temples, and pyramids that were made of limestone. The Mayas also carved important dates on tall stone blocks and used a kind of picture writing.

For unknown reasons, the Mayas abandoned their cities. When the Spaniards arrived in Guatemala in 1523, most of the Mayas were living in the Highlands.

The Spanish invader Pedro de Alvarado came to Guatemala from the Spanish colony in Mexico. Alvarado conquered all the major Indian groups in Guatemala and established Spanish rule.

On Sept. 15, 1821, Guatemala and other Central American states declared their independence. The states later became part of Mexico, but they broke away in 1823 and formed the United Provinces of Central America. The union established civil rights for the people and tried to curb the power of rich landowners and the Roman Catholic Church. However, conservative wealthy people and church officials worked against the union to regain their powers, and Guatemala withdrew from the union in 1839. The country then came under the rule of the first of its many dictators, Rafael Carrera.

Carrera was a conservative army general who restored privileges to the wealthy. After his death in 1865, a number of liberal presidents ruled Guatemala, but they were dictators too. They promoted economic development, especially coffee production, and supported immigration and investment by foreigners.

The Temple of the Giant Jaguar rises 150 feet (45 meters) at Tikal. These ruins are all that remains of a great Mayan city and ceremonial center.

The Guatemalan people, still seeking political freedom, started a 10-year political and economic revolution in 1944. In 1945, a new constitution gave the people liberties they had never known. A free press developed, and political parties formed. Under President Juan José Arévalo, the government promoted education, health care, and labor unions.

In 1952, President Jacobo Arbenz Guzmán began taking privately owned land and giving it to landless peasants. When the government began to take land from the United Fruit Company, however, the U.S. government feared that Guatemala's government was being influenced by Communists. The United States supported a successful revolt against Arbenz in 1954, and a temporary military government was set up.

Stone figures of Mayan gods mark ancient ceremonial sites. The Mayas worshiped gods and goddesses that influenced daily life, such as the corn god and the moon goddess. During their ceremonies, the Mayas sacrificed animals and sometimes humans.

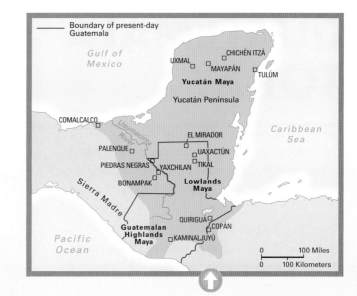

The land of the Maya included parts of present-day Guatemala as well as parts of Mexico, Honduras, and El Salvador, as well as all of Belize. This map shows the areas inhabited by three chief Maya groups and the location of major Maya cities.

Political confusion marked Guatemala from the 1960's through the mid-1980's. Between 1970 and 1982, four presidential elections were held, but many people claimed they were dishonest. Sometimes military dictators were elected; other times they seized power. In the late 1970's, violence became widespread when various rebel groups of *leftists*—people with socialist or Communist beliefs—fought the government. These rebel groups included poor rural people with little economic or political power.

In 1986, a civilian government was finally elected after one general had overthrown another. A new constitution was written, and the Congress was reestablished.

A new civilian president, Jorge Serrano Elías, was elected in 1991. In May 1993, Serrano suspended the Constitution, the Supreme Court, and the Constitutional Court. He also placed news organizations under censorship. The military promptly forced Serrano out of office. Days later, democratic rule was restored when Ramiro de Leon Carpio, a human rights advocate, was sworn in as president. Since then, voters have elected the country's presidents.

The violence that had begun in the late 1970's when various groups of leftists fought government forces for economic and political power ended in 1996 when the opposition groups and the government signed a peace agreement.

A colorful Spanish Baroque church dominates a Guatemalan town. About 70 per cent of the country's people are Roman Catholics.

GUATEMALA TODAY

Guatemala is the most populated of the Central American nations. Most of its people live in Guatemala's rugged central highlands. Guatemala City, the nation's capital and largest city, lies in this region.

The land and farming

The land of Guatemala is divided into three main regions. The Northern Plain, also called El Petén, is an area of thick tropical rain forests and some grasslands. The region gets from 80 to 150 inches (200 to 381 centimeters) of rain a year, and the temperature averages 80° F (27° C) the year around. Few people live in this area now, but many ancient Mayan ruins still stand in the forests. The country's largest lake, the 228-square-mile (591-square-kilometer) Lake Izabal, lies near the eastern Caribbean coast. Bananas are grown in this area.

The Highlands are a chain of mountains that stretch across Guatemala from east to west. This region has many volcanoes as well as the highest mountain in Central America—Volcán Tajumulco, which rises 13,845 feet (4,220 meters) above sea level. Earthquakes sometimes shake the region.

Most Guatemalans live in the Highlands because the soil is rich and the climate is mild. Mountain valleys have yearly average temperatures of 60° to 70° F (16° to 21° C). Most of the coffee-growing and corn-growing farmland is in the Highlands too. A great deal of coffee is exported, and corn is Guatemala's basic food crop. Some workers live on the coffee plantations that lie on the southern edge of the Highlands.

The Pacific Lowland runs along the coast between the ocean and the Highlands. The region is hot and humid and thinly populated. Many forest-lined streams flow from the mountains through the lowland. The production of cotton for export has become a major economic activity in the region. Farmers there also grow corn, rubber-bearing trees, and sugar cane, and raise beef cattle.

FACTS

Official name:	Republica de Guatemala (Republic of Guatemala)
Capital:	Guatemala City
Terrain:	Mostly mountains with narrow coastal plains and rolling limestone plateau
Area:	42,042 mi² (108,889 km²)
Climate:	Tropical; hot, humid in lowlands; cooler in highlands
Main rivers:	Motagua, Negro, Pasión, San Pedro
Highest elevation:	Volcán Tajumulco, 13,845 ft (4,220 m)
Lowest elevation:	Pacific Ocean, sea level
Form of government:	Republic
Head of state:	President
Head of government:	President
Administrative areas:	22 departamentos (departments)
Legislature:	Congreso de la Republica (Congress of the Republic) with 158 members serving four-year terms
Court system:	Corte Suprema de Justicia (Supreme Court), Court of Constitutionality
Armed forces:	15,500 troops
National holiday:	Independence Day - September 15 (1821)
Estimated 2010 population:	14,368,000
Population density:	42 persons per mi² (132 per km²)
Population distribution:	52% rural, 48% urban
Life expectancy in years:	Male, 67; female, 73
Doctors per 1,000 people:	0.9
Birth rate per 1,000:	33
Death rate per 1,000:	6
Infant mortality:	29 deaths per 1,000 live births
Age structure:	0-14: 43%; 15-64: 53%; 65 and over: 4%
Internet users per 100 people:	10
Internet code:	.gt
Languages spoken:	Spanish (official), Amerindian languages (more than 20 Amerindian languages, including Quiche, Cakchiquel, Kekchi, Mam, Garifuna, and Xinca)
Religions:	Roman Catholic, Protestant, indigenous Mayan beliefs
Currency:	Quetzal
Gross domestic product (GDP) in 2008:	$38.96 billion U.S.
Real annual growth rate (2008):	3.8%
GDP per capita (2008):	$2,879 U.S.
Goods exported:	Bananas, cardamom, clothing, coffee, crude oil, sugar
Goods imported:	Iron and steel, machinery, motor vehicles, petroleum products, ceutical products
Trading partners:	China, Costa Rica, El Salvador, Honduras, Mexico, United States

North

EL MIRADOR

Dos Lagunas

MIRADOR-DOS LAGUNAS-RÍO AZUL
NATIONAL PARK

Laguna del Tigre

Carmelita

UAXACTÚN

Paso Caballos

PIEDRAS NEGRAS

TIKAL
NATIONAL PARK

TIKAL

Melchor
de Mencos

BELIZE

SIERRA DE LACANDÓN
NATIONAL PARK

San Pedro

Lago
Petén Itzá

Flores

MEXICO

La Libertad

Santa Ana

Caribbean
Sea

Pasión

Sayaxché

Dolores

CEIBAL N.P.

ALTAR DE SACRIFICIOS

Poptún

Machaquilá

San Luis

Cancuén

Gulf of
Honduras

Bay of
Amatique

Cabo de
Tres Puntas

LAGUNA LACHUA
N.P.

Barillas

Chixoy

Sarstún

Livingston

Santa Cruz

Puerto
Barrios

Jacaltenango

Chahal

Salegua

Chahón

SIERRA DE SANTA CRUZ

RÍO DULCE
N.P.

SIERRA DE LOS CUCHUMATANES

Cobán

San Pedro Carchá

El Estor

Morales

L. Izabal

MONTAÑAS DEL MICO

Nebaj

Chixoy

Negro

ZACULEU

Volcán Tacaná
13,428 ft (4,093 m)

Huehuetenango

SIERRA DE CHUACÚS

Salamá

SIERRA DE LAS MINAS

QUIRIGUÁ

Gualán

Los Amates

Volcán Tajumulco
13,845 ft (4,220 m)

San Pedro Sacatepéquez

Santa Cruz del Quiché

Motagua

Zacapa

San Marcos

Totonicapán

Chichicastenango

HONDURAS

SIERRA

Quesaltenango

MIXCO VIEJO

El Progreso

Chiquimula

Pajapita

Sololá

Comalapa

Chimaltenango

Jalapa

San Luis Jilotepeque

Coatepeque

KAMINALJUYÚ

Esquipulas

ATITLÁN N.P.

Antigua

Guatemala City

Mazatenango

Amatitlán

Monte Cristo
7,933 ft (2,418 m)

Retalhuleu

Volcán de Acatenango
13,045 ft (3,976 m)

L. Ayarza

Asunción Mita

Santa Lucía Cotzumalguapa

El Progreso

Champerico

Tiquisate

Escuintla

Cuilapa

Jutiapa

Jalpatagua

PAN AMERICAN
HIGHWAY

Masagua

Semillero

La Gomera

Taxisco

Chiquimulilla

Sipacate

San José

Iztapa

SIPACATE-
NARANJO N.P.

Buena Vista

EL SALVADOR

Pacific Ocean

| 0 | | 50 | | 100 Miles |
| 0 | 50 | | 100 | 150 Kilometers |

GUATEMALA

Guatemala is a developing Central American country.
Thick forests cover almost half of this tropical land.
Two blue stripes on its flag represent the Atlantic and
Pacific oceans, which lie to the east and southwest.

The cone-shaped Tolimán is one
of the inactive volcanoes that
rise near beautiful Lago de
Atitlán. The lake bed is believed
to be an ancient valley that was
dammed by volcanic ash.

The cities and manufacturing

Many rural people have moved to Guatemala's cities,
and more keep coming. Guatemala City is one of the
largest cities in Central America. The nation's manufacturing industry is growing, but it cannot keep up with
urban population growth. As a result, the cities have
much unemployment. Guatemala's manufacturers produce mainly consumer goods, such as processed foods,
beverages, and clothes.

Economy

Agriculture is Guatemala's leading goods-producing industry. The nation's economy depends heavily on the
export of farm products.

In the early 1990's, Guatemala faced serious economic problems. Rising *inflation* (sharp increases in
prices) dramatically reduced workers' purchasing power.

Jorge Serrano Elías seized control of the government
during the economic turmoil of 1993. But Serrano's action drew international criticism and a halt to U.S. aid.
The move further threatened the economy, which was
heavily dependent on foreign aid. The crisis ended after
the military ousted Serrano and Congress elected
Ramiro de Leon Carpio as president.

PEOPLE

Like many other Latin American countries, Guatemala was colonized by Spain. Today, most of Guatemala's people belong to one of two groups—Indians or people of mixed Indian and European descent. In Guatemala, people of mixed descent are called *Ladinos*.

The Indians

Almost half the people of Guatemala—about 45 percent—are Indians. Their ancestors, the Maya, built a highly developed civilization hundreds of years before Europeans came to the Americas. Today, the Indians live in small peasant communities apart from the mainstream of Guatemalan life. Their ways differ greatly from those of other Guatemalans.

In Guatemala, as in Mexico, being called an Indian depends more on an individual's way of life and self-image than on the person's race. A Guatemalan is considered an Indian if that individual speaks an Indian language, wears Indian clothing, and lives in a community that follows the Indian way of life.

The Indians think of themselves more as part of their own small community than as part of the country of Guatemala. And there is little sense of political or social unity among the various Indian communities. Almost every Indian community has its own colorful style of clothing. Indians often travel far from home to trade in local markets or to find work. Most of them are extremely poor, and few are able to read and write. Almost all of Guatemala's Indians speak one of the many Mayan languages. In addition, many Indian men and some women speak Spanish.

Although most Indians are considered Roman Catholic, they also follow many of the religious practices of their ancestors. They worship local gods and spirits along with the Christian God, the Virgin Mary, and the saints. The Indians believe that their gods are present in nature, and they pray to them, especially during planting and harvesting time.

An Indian mother and child dress in their finest clothes for a religious festival in the village of Zunil. Such holidays provide relief from the hardships of everyday life.

Village women, the descendants of Maya Indians, chat on the step of an office in an Indian community. Today, Indians make up almost half the population of Guatemala.

Religious feast days provide the main sources of recreation among Indian peasants. On these holidays, the people hold processions, set off fireworks, and play marimba music. They also perform dances that tell stories from history or legends.

The Ladinos

About 55 percent of Guatemala's people are Ladinos, who follow Spanish American customs and traditions. Ladinos speak Spanish, the official language of the country. Again, whether a person is called a Ladino or an Indian does not depend entirely on that person's racial background. An Indian who drops the Indian way of life, speaks Spanish, and joins a Ladino community is considered a Ladino.

Some Ladinos are peasant farmers and laborers, particularly in the east and south, where few Indians live. The homes of Ladino peasants are simple, much like those of the Indians. Most live in one- or two-room houses made of adobe or poles lashed together, with palm, straw, or tile roofs. Their farm tools may include an ax, a digging stick, a hoe, and a machete. Farmers with flat, fertile land may have oxen and plows.

Ladinos make up the middle and upper classes in Guatemala, as they do in other Central American countries. Most Ladinos live in cities and towns. That is especially true in Indian areas, such as the western Highlands.

Ladinos control much of the economy and government of Guatemala. The customs and clothing of Ladinos vary little by region, but differ according to their income, occupation, and social class.

Guatemala's population is growing rapidly. The number of Ladinos is increasing much faster than the Indian population, largely because the Ladinos receive better health care in the cities. There has also been a slow shift among Indians toward the Ladino way of life.

In the Ladino cities, the people enjoy such sports as basketball, bicycling, and soccer. On religious feast days, Ladinos in rural areas celebrate as the Indians do.

A colorful flower market is held in the central square in the village of Chichicastenango. On market days, peasants bring their crops and hand-crafted wares, such as pottery, to sell or trade. Sunday is the main market day.

Residents of Antigua enjoy Sunday dancing in the city's Central Park. Music is provided by musicians playing a popular Latin American percussion instrument called a marimba.

GUINEA

Guinea is a West African country with a tropical climate and a variety of landscapes. A swampy coastal strip rises to the Fouta Djallon, a central plateau of hard, crusty soil. In the north lie grassy plains called *savannas,* while in the southeast the forested hills of the Guinea Highlands rise more than 5,000 feet (1,000 meters).

Mangrove trees line the mouths of Guinea's many rivers. Antelope, buffalo, crocodiles, elephants, hippopotamuses, leopards, lions, and monkeys are among the wildlife that make their home in the country.

The discovery of ancient stone tools has led scientists to believe that people have lived in the area of Guinea since prehistoric times. Hunters and then farmers inhabited the area.

Several powerful empires ruled parts of Guinea between 1000 and the mid-1400's. The Mali Empire, founded by the Malinke people, ruled the region during the 1200's. Later, the Songhai Empire conquered the area. During the 1600's, Muslim Fulani people from the north moved to Guinea and fought a *jihad* (holy war) against the Malinke for control of the Fouta Djallon.

The Portuguese were the first Europeans to reach Guinea, beginning in the mid-1400's. They captured and enslaved many Guineans. By the mid-1800's, France had gained some territory in the region through treaties and others through conquest. In 1891, Guinea became a French colony called French Guinea.

In 1947, a political party called the *Parti Démo-cratique de Guinée* (Democratic Party of Guinea), or PDG, was formed. Sékou Touré became head of the PDG in 1952, and by 1957, the party won control of the Guinean legislature. Guineans gained full independence from France in 1958, and Touré became the country's first president.

FACTS

Official name:	Republique de Guinee (Republic of Guinea)
Capital:	Conakry
Terrain:	Generally flat coastal plain, hilly to mountainous interior
Area:	94,926 mi² (245,857 km²)
Climate:	Generally hot and humid; monsoonal-type rainy season (June to November) with southwesterly winds; dry season (December to May) with northeasterly harmattan wind
Main rivers:	Niger, Konkouré, Gambia
Highest elevation:	Mont Nimba, 5,748 ft (1,752 m)
Lowest elevation:	Atlantic Ocean, sea level
Form of government:	Republic
Head of state:	President
Head of government:	Prime minister
Administrative areas:	7 regions administrative (administrative regions), 1 zone speciale (special zone)
Legislature:	Assemblee Nationale Populaire (People's National Assembly) with 114 members serving five-year terms
Court system:	Cour d'Appel (Court of Appeal)
Armed forces:	12,300 troops
National holiday:	Independence Day - October 2 (1958)
Estimated 2010 population:	10,088,000
Population density:	106 persons per mi² (41 per km²)
Population distribution:	67% rural, 33% urban
Life expectancy in years:	Male, 54; female, 57
Doctors per 1,000 people:	0.1
Birth rate per 1,000:	40
Death rate per 1,000:	12
Infant mortality:	93 deaths per 1,000 live births
Age structure:	0-14: 43%; 15-64: 54%; 65 and over: 3%
Internet users per 100 people:	0.9
Internet code:	.gn
Languages spoken:	French (official), African languages
Religions:	Muslim 85%, Christian 8%, indigenous beliefs 7%
Currency:	Guinean franc
Gross domestic product (GDP) in 2008:	$4.42 billion U.S.
Real annual growth rate (2008):	2.9%
GDP per capita (2008):	$440 U.S.
Goods exported:	Alumina, bauxite, coffee, diamonds, fish, gold
Goods imported:	Machinery, motor vehicles, petroleum products, pharmaceuticals, rice and other food products
Trading partners:	China, France, Russia, Spain, Ukraine, United States

Under Touré and the PDG, the government took nearly complete control of the economy and tried to create a socialist state. Throughout the 1960's and early 1970's, Touré also crushed all opposition to his policies and threw many of his opponents into prison. By the late 1970's, Touré began to relax some of the government's political restrictions and release political prisoners. Economic problems remained, however.

Touré died in 1984, and army officers took control of the government. Colonel Lansana Conté then became president. His government abandoned the socialist policies of Touré and adopted free enterprise policies for the Guinean economy.

In 1990, voters approved a new constitution that provided for a return to civilian rule. In a multiparty election held in 1993, Conté was elected president. He was reelected in 1998 and 2003. After Conté died in 2008, a group of military leaders seized power in Guinea. A presidential election held in June 2010 was widely condemned as fraudulent. In a November runoff, opposition leader Alpha Condé was elected president.

Guinea, a small country on the western bulge of Africa, is a land of coastal swamps, plateaus, grassy plains, and forested hills. The capital, Conakry, is also the largest city in this underdeveloped nation.

A typical Guinean village is made up of round huts with mud walls and thatched roofs. Urban dwellers live in one-story rectangular houses made of mud bricks or wood.

PEOPLE AND ECONOMY

Almost all Guineans are black Africans, and about 75 percent of the people belong to one of three main ethnic groups. Most of the Fulani, or Peul—the largest group—live in the central plateau region called the Fouta Djallon. Members of the second largest group, the Malinke, live mainly in the northeastern section of the country, especially in the towns of Kankan, Kouroussa and Siguiri. Members of the third largest ethnic group, the Susu, live along the coast.

About two-thirds of all Guineans live in rural areas. They wear traditional clothing and, like their ancestors, live in round huts. In the cities and towns, however, some Guineans wear Western-style clothes. Most urban dwellers live in one-story rectangular houses made of mud bricks or wood. A serious housing shortage exists in the capital city of Conakry, however.

While French is the nation's official language, most Guineans speak one of eight African languages. The people have a rich popular culture that they express through folk tales, dramas, and music. History is recited by storytellers called *griots*.

Most Guinean adults cannot read and write. Public schools are free, and all children between the ages of 7 and 19 are required to attend. But few actually go to school, partly due to a shortage of teachers and classrooms. Many Guineans send their children to private Muslim schools.

Most of Guinea's people are farmers, but many raise barely enough food for their families. Rice is the main food of Guineans who live near the coast, while corn and millet are the basic foods of people on the northern savannas. These grains are usually pounded into meal, mixed with water, and boiled into a porridge that is served with a hot, spicy sauce. Fruits, such as bananas, pineapples, and plantains, and such vegetables as cassava are sometimes included in meals. Meat and fish are served occasionally, but the diet of most Guineans lacks protein and vitamins.

Thatched huts and a single tin-roofed building tucked amid tropical vegetation make up a rural settlement in the forested hills of the Guinea Highlands. Most Guineans live in such small villages.

A herdswoman drives her cattle along a rural road. Some farmers in the plains and highlands of Guinea raise livestock for a living. Guineans occasionally eat meat, and some drink milk mixed with water.

Guinean children sit on a boxcar carrying bauxite, the country's major source of wealth. The mineral is used to make aluminum. Bauxite and its processed form (alumina), together with gold, make up the majority of Guinea's export income.

Workers unload rice in Conakry, one of Guinea's two international shipping ports. Rice is the main food for many Guineans. Although farmers grow enough rice to feed their own families, the nation must still import rice to meet all its people's needs.

Balancing loaves of bread on her head, a young woman threads her way through automobile traffic in Conakry. Poor transportation has limited Guinea's economic development.

Although Guinea is one of the least developed nations in the world, it has many natural resources that could make it prosperous one day. For example, Guinea has about one-third of the world's reserves of bauxite, a mineral used to make aluminum. Other important mineral resources include deposits of iron ore, diamonds, gold, and uranium.

Mining, manufacturing, and construction together employ about 10 percent of Guinean workers. Factory workers manufacture food products and textiles and refine bauxite to produce aluminum. Craftworkers create woven baskets, metal jewelry, and leather goods.

Economic development in Guinea is also hampered by the country's poor transportation systems. The roads, usually unpaved, are in poor condition. Most of the country's railroad tracks also need repair. An international airport operates in Conakry. Conakry and Kamsar serve as international shipping ports.

Communications are also limited in Guinea. The government controls the country's newspapers, radio, and television. Most Guineans own radios, but few have television sets or computers.

GUINEA-BISSAU

The republic of Guinea-Bissau is a tiny West African nation wedged between Senegal to the north and Guinea to the south and east. Many rivers flow through this very warm, rainy land, which slopes upward from a forested and swampy coast to inland grasslands called *savannas*.

Most of the people of Guinea-Bissau are farmers. They live in straw huts with thatched roofs, and they grow beans, coconuts, corn, palm kernels, peanuts, and rice. Only a small percentage of the people work in the country's few industries—mainly in building construction and food processing.

About 85 percent of Guinea-Bissauans are black Africans who belong to about 20 different ethnic groups, including the Balanta, Fulani, Manjako, and Mandinka—the largest groups. The rest of the population are *mulattoes,* people of mixed black African and Portuguese ancestry.

Although education has improved since the country won its independence in 1974, only about a third of all adults in Guinea-Bissau can read and write. The law requires all children from the ages of 7 to 13 to attend school, but only about half actually do so.

Many ethnic groups of black Africans were already living in what is now Guinea-Bissau when the Portuguese explorers arrived in the region in 1446. From the 1600's to the 1800's, the Portuguese used the area as a base for their slave trade. It became a Portuguese colony called Portuguese Guinea in 1879, and an overseas province in 1951.

In 1956, African nationalist leaders founded the African Party for the Independence of Guinea and Cape Verde—often called the PAIGC, the party's initials in Portuguese. The party, headed by Amilcar Cabral, sought independence for both Portuguese Guinea and the island group called Cape Verde. During the early 1960's, the PAIGC trained many farmers in the hit-and-run tactics of guerrilla warfare. It also established many schools and adult education programs.

A war for independence began in 1963, and by 1968, the PAIGC controlled about two-thirds of the province. The people in these areas elected the first National Popu-

FACTS

Official name:	**Republica da Guine-Bissau (Republic of Guinea-Bissau)**
Capital:	**Bissau**
Terrain:	**Mostly low coastal plain rising to savanna in east**
Area:	**13,948 mi² (36,125 km²)**
Climate:	**Tropical; generally hot and humid; monsoonal-type rainy season (June to November) with southwesterly winds; dry season (December to May) with northeasterly harmattan winds**
Main rivers:	**Cacheu, Corubal, Geba**
Highest elevation:	**Unnamed location in the southeastern part of the country, 984 ft (300 m)**
Lowest elevation:	**Atlantic Ocean, sea level**
Form of government:	**Republic**
Head of state:	**President**
Head of government:	**Prime minister**
Administrative areas:	**9 regioes (regions)**
Legislature:	**Assembleia Nacional Popular (National People's Assembly) with 102 members serving a maximum of four years**
Court system:	**Supremo Tribunal da Justica (Supreme Court), Regional Courts, Sectoral Courts**
Armed forces:	**6,500 troops**
National holiday:	**Independence Day - September 24 (1973)**
Estimated 2010 population:	1,803,000
Population density:	29 persons per mi² (50 per km²)
Population distribution:	70% rural, 30% urban
Life expectancy in years:	Male, 45; female, 48
Doctors per 1,000 people:	0.1
Birth rate per 1,000:	47
Death rate per 1,000:	18
Infant mortality:	117 deaths per 1,000 live births
Age structure:	0-14: 48%; 15-64: 49%; 65 and over: 3%
Internet users per 100 people:	2
Internet code:	.gw
Languages spoken:	Portuguese (official), Crioulo, African languages
Religions:	Indigenous beliefs 50%, Muslim 40%, Christian 10%
Currency:	Communaute Financiere Africaine franc
Gross domestic product (GDP) in 2008:	$444 million U.S.
Real annual growth rate (2008):	3.2%
GDP per capita (2008):	$306 U.S.
Goods exported:	Mostly: cashew nuts Also: fish products, sawn lumber
Goods imported:	Food, machinery, motor vehicles, petroleum products
Trading partners:	India, Nigeria, Portugal, Senegal

lar Assembly in 1972, and in 1973, the Assembly declared the province to be an independent nation called Guinea-Bissau. Amilcar Cabral was assassinated, but his brother, Luis Cabral, became the new nation's first president. The war ended in 1974, when Portugal recognized Guinea-Bissau's independence. Cape Verde became independent in 1975.

The government then tried to increase farm production. During the war, many crops were destroyed. Government leaders planned to farm unused land and modernize farming methods. They also wanted to develop the nation's mineral resources, but because of political instability and a shortage of skilled workers, these plans had little success.

The PAIGC worked to unite Guinea-Bissau and Cape Verde under one government. But in 1980, army officers who opposed such a union overthrew Guinea-Bissau's government and abolished the Assembly. In 1984, a new constitution was adopted, and a new National Assembly was established. Brigadier General João Bernardo Vieira became president. The country held its first multiparty elections in 1994.

During the late 1990's and the first decade of the 2000's, political violence again became commonplace in Guinea-Bissau. Vieira, who had been reelected president in 2005, was killed in 2009. Even after elections to replace him, tension between military and government leaders contributed to cause unrest.

A street vendor waits for customers on the main road in Bissau, the capital of Guinea-Bissau. The country's economy relies primarily on agricultural products.

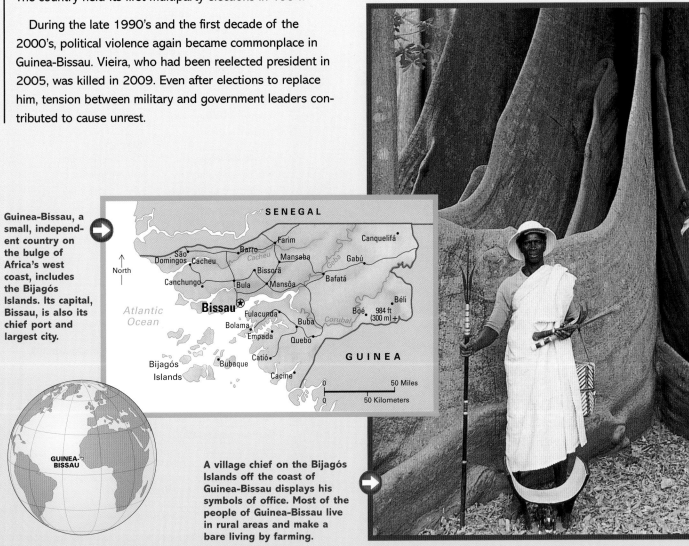

Guinea-Bissau, a small, independent country on the bulge of Africa's west coast, includes the Bijagós Islands. Its capital, Bissau, is also its chief port and largest city.

A village chief on the Bijagós Islands off the coast of Guinea-Bissau displays his symbols of office. Most of the people of Guinea-Bissau live in rural areas and make a bare living by farming.

GUYANA

The republic of Guyana, on the northeast coast of South America, is a tropical land of sugar cane plantations and rice farms. Rain forests and mountains cover much of the interior of the country, making most of Guyana difficult to reach. Isolated from the rest of South America by its terrain, Guyana more closely resembles the islands of the Caribbean in its culture, history, and economy.

Guyana is an American Indian word meaning *Land of Waters*. The country's official name is the Cooperative Republic of Guyana. Guyana is made up of people from several national and racial groups. East Indians and blacks form the largest groups.

Early days

European explorers first arrived in Guyana during the late 1500's and early 1600's. They found Arawak, Carib, and Warrau Indians living in the area. In 1581, the Dutch founded a settlement in what is now Guyana and claimed the area. Later, the British and the French also claimed it. The United Kingdom gained control of the land in 1814 and formed the colony of British Guiana in 1831.

The early Dutch settlers set up sugar cane plantations, and the plantation owners brought in black Africans to work as slaves. When slavery was made illegal in the 1830's, many of the newly freed blacks went to live in the towns. Plantation owners then began to import laborers from India.

During the 1940's and 1950's, the British became more active in preparing the colony for self-government. More of the people were allowed to vote, and more members of the legislature were elected by the people.

Independence and after

In 1953, a new constitution was adopted, and British Guiana held its first election based on *universal suffrage* (the right of all adults to vote). The People's Progressive Party (PPP), led by Cheddi B. Jagan, won most of the seats in the legislature. The British government, however, suspended the Constitution, believing that Jagan's administration would turn British Guiana into a Communist state.

FACTS

Official name:	Cooperative Republic of Guyana
Capital:	Georgetown
Terrain:	Mostly rolling highlands; low coastal plain; savanna in south
Area:	83,000 mi^2 (214,969 km^2)
Climate:	Tropical; hot, humid, moderated by northeast trade winds; two rainy seasons (May to mid-August, mid-November to mid-January)
Main rivers:	Demerara, Essequibo, Courantyne, Berbice
Highest elevation:	Mount Roraima, 9,094 ft (2,772 m)
Lowest elevation:	Atlantic Ocean, sea level
Form of government:	Republic
Head of state:	President
Head of government:	Prime minister
Administrative areas:	10 regions
Legislature:	National Assembly with 65 members serving five-year terms
Court system:	Supreme Court of Judicature
Armed forces:	1,100 troops
National holiday:	Republic Day - February 23 (1970)
Estimated 2010 population:	757,000
Population density:	9 persons per mi^2 (4 per km^2)
Population distribution:	71% rural, 29% urban
Life expectancy in years:	Male, 64; female, 69
Doctors per 1,000 people:	0.5
Birth rate per 1,000:	19
Death rate per 1,000:	8
Infant mortality:	39 deaths per 1,000 live births
Age structure:	0-14: 29%; 15-64: 66%; 65 and over: 5%
Internet users per 100 people:	25
Internet code:	.gy
Languages spoken:	English, Amerindian dialects, Creole, Hindi, Urdu
Religions:	Christian 57%, Hindu 28%, Muslim 10%, other 5%
Currency:	Guyanese dollar
Gross domestic product (GDP) in 2008:	$1.14 billion U.S.
Real annual growth rate (2008):	3.2%
GDP per capita (2008):	$1,515 U.S.
Goods exported:	Bauxite, diamonds, gold, rice, shrimp, sugar, timber
Goods imported:	Food, machinery, motor vehicles, petroleum products
Trading partners:	Canada, Trinidad and Tobago, United Kingdom, United States

The republic of Guyana lies nestled between Venezuela, Brazil, and Suriname. Guyana is an American Indian word meaning Land of Waters. The region was one of the first in the Western Hemisphere to be settled. Christopher Columbus sailed along its coast in 1498, and Sir Walter Raleigh arrived in 1598 in search of the legendary El Dorado. Guyana became independent in 1966.

A butcher in the capital city of Georgetown displays meats that comply with the dietary requirements of the nation's Muslims. East Indians—both Muslim and Hindu—make up about 50 percent of the population. They are descendants of people brought from India in the 1830's to work on plantations.

In 1955, the PPP split apart, and in 1957, Jagan's deputy, Forbes Burnham, founded the People's National Congress (PNC). The PNC attracted the support of the black population, while the East Indians favored the PPP.

British Guiana became the independent nation of Guyana in 1966, and Forbes Burnham became the country's first prime minister. From 1968 to the early 1990's, the PNC held the majority of seats in the nation's legislature, and as a result, blacks held most of the leading positions in the government and armed forces. However, East Indians continued to dominate in rice production and commerce.

Burnham ruled as prime minister, and later as president, until he died in 1985. His successor, Hugh Desmond Hoyte, introduced economic reforms. Many state-owned companies were sold to private investors.

In 1992 elections, the PPP won a majority and Cheddi B. Jagan became president. He died in 1997, and his widow, American-born Janet Jagan, was elected president. She was succeeded in 1999 by Bharrat Jagdeo, also of the PPP. Jagdeo was reelected in 2001 and 2006. In the 2011 election, the PPP lost its majority, and Donald Ramotar succeeded Jagdeo as president.

LAND AND PEOPLE

Much of Guyana's coastal region, which lies along the North Atlantic Ocean, was swampland until the early European settlers drained it. This coastal strip, which is only 2 miles (3.2 kilometers) wide in some places and only about 30 miles (48 kilometers) at its widest point, is home to 90 percent of the country's population.

Rice fields and sugar cane plantations cover the coastal plains. *Dikes* (dams built to prevent flooding), sea walls, and a system of drainage canals prevent the sea from flooding the land and protect people and their crops.

South of the coastal plain lies the inland forest region, which covers about 85 percent of Guyana. The area contains about a thousand different types of timber, as well as many plant and animal species. This vast region is almost uninhabited.

Beyond the forest lie the highland areas, which consist of mountains and *savannas* (high treeless plains). The Rupununi savanna stretches along the Brazilian border in the southwest, while a smaller savanna lies in the northeast. Like the forests, the savannas are largely uninhabited, except for a few surviving tribes of *Amerindians* (American Indians) who lead a primitive life almost completely isolated from the rest of the population.

Rivers and waterfalls

Guyana's four main rivers—the Essequibo, the Demerara, the Berbice, and the Courantyne—flow northward from the savannas through the rain forests before emptying into the Atlantic Ocean. Heavy rainfall in the forests often causes the rivers to swell, and spectacular waterfalls and rapids can be found along their courses.

The most spectacular waterfall in Guyana is the Great Fall, which drops 1,600 feet (488 meters) on the Kamarang River. Great Falls drops 840 feet (256 meters) on the Mazaruni River, and Kaieteur Fall on the Potaro River drops 741 feet (226 meters).

A timber-framed building in British colonial style is typical of the architecture of Georgetown, the capital and chief city of Guyana. Georgetown, on the east bank of the Demerara River facing the Atlantic Ocean, is the main outlet for Guyana's exports.

An aerial view of Guyana's landscape shows the country's major land regions and geographical features. Here, a river flows in a winding course through mountains, grassy savannas, and rain forests.

Canal barges carry sugar cane, a major cash crop, to a factory for processing. An extensive canal network, along with sea walls and dikes, protects Guyana's low-lying coastal plains against floods.

Guyana's wildlife includes the giant otter (1) and the hoatzin (2), a brightly colored bird that lives in the marshy areas. Hoatzins are born with claws on their wings (3), which they use to climb on tree branches until they learn to fly. The claws fall off as the birds mature.

Guyana's rivers—often the only route through the dense forests of the interior—provide an important means of transportation and communication. Farther north, the rivers serve as shipping routes for the mining industry. Although the town of Linden lies about 65 miles (105 kilometers) inland, it is Guyana's most important mining center because the Demerara River is wide enough at that point for ocean-going ships.

People

Guyana's East Indians—descendants of people who were brought to work on the sugar plantations—make up the nation's largest ethnic group. They account for about half of the population. Guyana's second largest ethnic group consists of blacks whose ancestors were brought from Africa as slaves. Tension between the East Indians and blacks caused much political turmoil during the mid-1900's.

Most East Indians live in the rural areas and work on sugar plantations or on small farms. However, increasing numbers of East Indians have moved to the cities and become merchants, doctors, and lawyers. Blacks live in cities and towns and often work as teachers, police officers, government employees, and skilled workers in the sugar-grinding mills and bauxite mines.

English is Guyana's official language, and it is spoken by most of the people. Many of the people speak a regional form of English called Creole. Hindi and Urdu are widely used by Guyana's East Indians.

HAITI

Haiti is a small country on the western third of the island of Hispaniola, which lies in the Caribbean Sea between Cuba and Puerto Rico. It is one of the most densely populated countries in the Western Hemisphere, and one of the least developed.

Founded in 1804, Haiti is the oldest black republic in the world, and the second oldest independent nation in the West. Only the United States is older.

The land

Two rugged mountain chains run across Haiti, one in the north and one in the south. Each forms a peninsula at the western end of the country. The northern peninsula extends about 100 miles (160 kilometers) into the Atlantic Ocean, and the southern peninsula juts about 200 miles (320 kilometers) into the Caribbean Sea. Tropical pine and mahogany forests cover some of the mountains, while tropical fruit trees grow on others.

A gulf, the Golfe de la Gonâve, separates the western peninsulas, and an island, Île de la Gonâve, lies within the gulf. The broad valley of the Artibonite River runs between the mountains in the eastern part of the country.

Haiti has a mild tropical climate. Destructive hurricanes sometimes strike the country between June and October.

History

Christopher Columbus landed on the island of Hispaniola in 1492 and claimed it for Spain. Spanish settlers soon arrived and eventually caused the deaths of almost the entire Arawak Indian population. Many Spaniards later left Hispaniola for more prosperous settlements, and pirates took over the western coast. In 1697, Spain recognized French control of that part of the island.

France named its new colony Saint Domingue. Slaves were brought from Africa to work on plantations. In 1791, the slaves rebelled, and, in 1803, they defeated the French army. On Jan. 1, 1804, the rebel leader General Jean Jacques Dessalines proclaimed the land an independent country named Haiti.

After Dessalines's death in 1806, a decades-long power struggle began. From 1844 to 1914, 32 different men ruled the nation. U.S. troops occupied Haiti from 1915 to 1934.

FACTS

Official name:	Republique d'Haiti (Republic of Haiti)
Capital:	Port-au-Prince
Terrain:	Mostly rough and mountainous
Area:	10,714 mi² (27,750 km²)
Climate:	Tropical; semiarid where mountains in east cut off trade winds
Main rivers:	Artibonite, Les Trois Rivieres
Highest elevation:	Pic La Selle, 8,783 ft (2,677 m)
Lowest elevation:	Caribbean Sea, sea level
Form of government:	Elected government
Head of state:	President
Head of government:	Prime minister
Administrative areas:	10 departements (departments)
Legislature:	Assemblee Nationale (National Assembly) consisting of the Senate with 30 members serving six-year terms and the Chamber of Deputies with 99 members serving four-year terms
Court system:	Cour de Cassation (Supreme Court)
Armed forces:	None
National holiday:	Independence Day - January 1 (1804)
Estimated 2010 population:	9,723,000
Population density:	908 persons per mi² (350 per km²)
Population distribution:	57% rural, 43% urban
Life expectancy in years:	Male, 58; female, 61
Doctors per 1,000 people:	0.3
Birth rate per 1,000:	29
Death rate per 1,000:	9
Infant mortality:	57 deaths per 1,000 live births
Age structure:	0-14: 38%; 15-64: 58%; 65 and over: 4%
Internet users per 100 people:	11
Internet code:	.ht
Languages spoken:	French (official), Creole (official)
Religions:	Roman Catholic 55%, Protestant 28%, other 17% (Note: many Christians also practice voodoo and consider Christianity their primary religion)
Currency:	Gourde
Gross domestic product (GDP) in 2008:	$6.96 billion U.S.
Real annual growth rate (2008):	2.3%
GDP per capita (2008):	$770 U.S.
Goods exported:	Clothing, cocoa, fruit, oils
Goods imported:	Machinery, motor vehicles, petroleum products, rice, sugar
Trading partners:	Dominican Republic, Netherland Antilles, United States

François Duvalier was elected president in 1957. In 1964, he declared himself president for life and ruled as dictator. His son, Jean-Claude Duvalier, succeeded him in 1971. Both men employed violent secret police called the *Tontons Macoutes* (bogeymen).

In 1986, Haitians overthrew Duvalier. The next three leaders of Haiti, including Jean-Bertrand Aristide, who was elected president in 1990, were also overthrown. In 1994, the Security Council of the United Nations (UN), citing human rights abuses and the mass exodus of Haitian refugees, authorized a U.S.-led invasion of Haiti to oust the military regime and reinstate Aristide. A negotiated settlement allowed U.S. troops to occupy the country without resistance and return power to Aristide. From 1995 to 1998, UN peacekeepers kept order in Haiti.

Aristide was again elected president in 2000. In 2004, he resigned during an armed rebellion, and U.S. forces maintained order in the country until the arrival of UN peacekeepers. Since that time, natural disasters, such as hurricanes and tropical storms, and a poor economy caused suffering and continued political instability in Haiti. In 2010, a powerful earthquake destroyed the capital, Port-au-Prince. About 316,000 people were killed and more than 1 million others were left homeless. Voters in 2011 elected popular Haitian singer Michel Martelly as president.

Deforested mountains lie near Jacmel in southern Haiti. Without trees to anchor the soil, erosion has reduced the country's scarce agricultural land and made the island more vulnerable to destructive floods each hurricane season.

Haiti is on the western end of the island of Hispaniola in the Caribbean Sea. It is the Western Hemisphere's poorest nation.

François "Papa Doc" Duvalier, a brutal dictator, controlled Haiti from 1957 to 1971. After his death, his son, Jean-Claude "Baby Doc" became president.

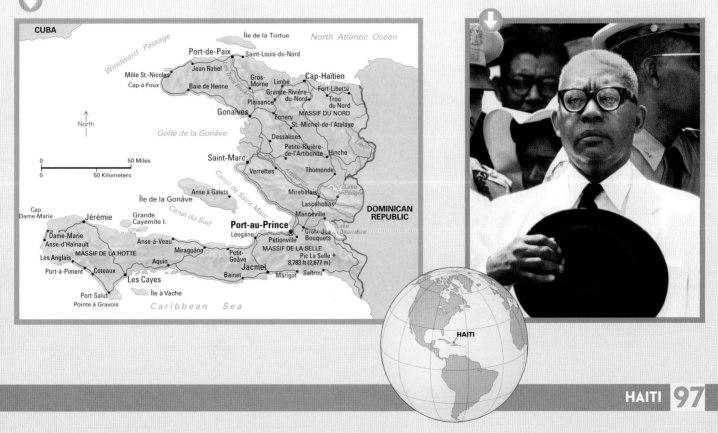

PEOPLE AND ECONOMY

Haiti is one of the most densely populated countries in the Western Hemisphere and also one of the poorest. About one-half of its people cannot read and write. Most are farmers who raise barely enough food to feed their families.

Ancestry

The majority of Haitians are descended from black Africans who were brought to Haiti to work as slaves. Most of these Haitians are crowded onto the country's coastal plains or in the mountain valleys.

About 5 percent of the people of Haiti are *mulattoes,* people of mixed African and European ancestry. Most of the mulattoes belong to the middle or upper class, and many have been educated in France. Most live in comfortable, modern homes. Some are prosperous merchants, doctors, or lawyers, and a few own large plantations. Haiti is also home to small populations of Americans, Europeans, and Syrians.

Middle- and upper-class Haitians speak French, the official language of the country. However, most Haitians speak a language called Haitian Creole, which is partly based on French.

Way of life

About one-fourth of the Haitian people are Protestants. But most Haitians are Roman Catholics, and many of them practice *voodoo.* Voodoo blends African and Christian beliefs. Voodoo followers believe in many gods, such as gods of rain, love, war, and farming. They also believe that they can be possessed by gods if they perform certain ceremonies. In one such ceremony, for example, a voodoo priest called a *houngan* draws a design on the ground with flour, and the people dance until they believe a god has possessed one or more of them.

Farmworkers harvest rice, above, in the Artibonite River Valley in central Haiti. Rice is a major food crop, but most farm families have barely enough land to raise rice for their own use.

A church in Port au Prince shows the effects of a devastating earthquake that struck Haiti on Jan. 12, 2010. The earthquake killed about 230,000 people and left more than 1 million homeless.

A local fisherman, carrying a fish trap made of split bamboo paddles his dugout canoe through the shallow, clear-blue waters off Haiti's coast.

A brightly colored vehicle—part bus, part truck—loads passengers and goods. Only a small percentage of Haiti's roads can be used in all kinds of weather.

More than half of Haiti's people live in rural areas, and most are farmers. A typical Haitian family farm is a small plot of land less than 2 acres (0.8 hectare) in size that was once part of the plantation where the family's slave ancestors worked.

In the 1970's, many Haitians began to leave their country because of poor economic conditions and political oppression. Following the 1991 coup, in which Aristide was overthrown, many Haitians attempted to flee to the United States in small boats. However, the U.S. government forced them to return to Haiti or sent them to the U.S. military base at Guantanamo, Cuba.

Agriculture

Most Haitian farmers raise barely enough food for themselves—mainly beans, corn, rice, and yams. If they are fortunate, they also have some chickens, a pig, or a goat. A typical family lives in a small, one-room hut with a thatched roof and walls made of sticks covered with dried mud.

The people farm as much of their land as they can. In some areas, they raise crops on slopes so steep that the farmers must anchor themselves with rope to keep from sliding down the hillside. Coffee, fruits, and *cacao*—a seed used to make chocolate—are grown in the mountains. The coffee beans and fruits are sold in the markets.

Most Haitians still follow some of their ancestors' African customs. Much of the work on the small farms of Haiti is done by groups of neighbors. They move from field to field, planting or harvesting crops to music and song in a combination of work and play called a *combite*.

On large plantations, laborers help raise coffee, sugar cane, or *sisal*, a plant used to make twine. Sugar cane is the main crop in the black, fertile soil of the Artibonite Valley. Many Haitians also work on plantations in the Dominican Republic or Cuba.

Haiti has few industries. Some factories process coffee and sugar cane for sale to the United States, France, and other countries. Haiti has a few cotton mills. Craft workers weave objects from sisal or carve figures out of mahogany. Tourism was once an important part of Haiti's economy, but it has decreased since the 1970's because of political unrest.

HONDURAS

Honduras is a small country in Central America. Centuries ago, the Mayas lived in what is now western Honduras and built the magnificent ceremonial center at Copán, with its beautiful palaces, pyramids, and temples. The Mayas studied astronomy and invented a calendar. They also developed a number system and picture writing. The Mayas played ball games too, and a Mayan ball court can still be seen at Copán. The center thrived until the 800's. But by the time Europeans arrived, Copán lay in ruins, and the Indians of the region had forgotten the city. Little else is known about Honduras before the 1500's.

Christopher Columbus was the first European to see Honduras. On his fourth voyage, he sailed to what are now the Bay Islands off the northern coast. Columbus landed at Cabo de Honduras (Cape Honduras) on July 30, 1502, and claimed the land for Spain.

A number of Spanish explorers soon visited the region and founded settlements. Many Indians were killed by the Spaniards or died of diseases brought by the Spanish colonists. Some Indians were shipped as slaves to plantations on the Caribbean islands; others worked in the gold and silver mines started by the Spaniards. The Spaniards also brought people from Africa to work as slaves in the mines. But the mines were never profitable enough to attract many colonists.

Honduras was a Spanish colony for about 300 years. On Sept. 15, 1821, Honduras and four other Central American states claimed independence. The states became part of Mexico for a short time, but in 1823 they broke away and formed the United Provinces of Central America. This union had liberal political and economic policies. It established civil rights and tried to curb the power of rich landowners and the Roman Catholic Church.

Honduras withdrew from the union in 1838, after the union began to fall apart under various pressures, including efforts by conservative wealthy people and church officials to regain their powers.

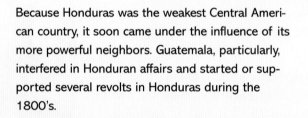

The base of a Maya pyramid stands at Copán, an eastern center of the Maya civilization that thrived until the 800's.

Because Honduras was the weakest Central American country, it soon came under the influence of its more powerful neighbors. Guatemala, particularly, interfered in Honduran affairs and started or supported several revolts in Honduras during the 1800's.

In the late 1800's, U.S. fruit companies began to arrive in Honduras, establishing banana plantations on the northern coast. Because of the income they brought to the country, the banana companies had a strong influence over the Honduran government. Honduras became known as a "banana republic" because of this relationship.

Until 1933, most Honduran presidents served short terms because they were overthrown in the nation's frequent revolutions. In that year, General Tiburcio Carías Andino became president and ruled as a dictator until 1948, despite several revolts.

During the 1950's, more political violence occurred. Then in 1957, Ramón Villeda Morales, a doctor, was elected president. Villeda started a land reform program and built hospitals, roads, and schools.

The Church of the Sorrows in Tegucigalpa was completed in 1732 in the Spanish Colonial style. On the front of the church are figures representing the passion of Jesus Christ.

A Maya head stands among the ruins at Copán. The ancient city featured beautiful stone palaces, pyramids, and temples.

In 1963, however, the government was overthrown by Colonel Osvaldo López Arellano and other army officers. A new constitution was written, allowing López to become president. In 1971, voters again elected a civilian to the presidency. But another military revolt followed, and López returned to power.

López himself was overthrown in 1975. Two more military-led governments followed. In the 1980's, however, the people elected a civilian president and also chose a new legislature. Since then, voters have elected a new president and legislature every four years.

During the 1980's, Honduras experienced problems with its neighbor, Nicaragua. Rebels fighting the Nicaraguan government crossed into Honduras and established bases from which they raided Nicaragua. Nicaraguan government troops sometimes entered Honduras to attack these rebels. In 1988, the rebels and the government signed a cease-fire agreement. In 1990, the Nicaraguan government was voted out of office, and the rebel bases in Honduras were shut down.

In 1998, Hurricane Mitch caused great destruction in Honduras. It destroyed thousands of homes and half of the country's crops.

HONDURAS TODAY

Honduras was named by an explorer—perhaps Columbus—for the deep waters off its northern coast. The Spanish word *honduras* means *depths*. Today, the country is known for the bananas it produces.

The land

Honduras has four main land regions. The Northern Coast is the banana-producing region of the country. Bananas are especially important in the fertile Ulua-Chamelecón River Basin and along the coastal plain near the port of Tela. East of Tela, the region is largely undeveloped and sparsely populated. Grasslands, swamps, and forests cover the hot, humid land.

The Southern Coast on the Gulf of Fonseca, a small arm of the Pacific Ocean, is lined with mangrove trees. Narrow plains lie just inland. The largest of these plains, along the Choluteca River, is a fertile area of farms and cattle ranches. The Southern Coast is hot and humid, but receives less rain than the Northern Coast.

The Mountainous Interior covers more than 60 percent of the nation. One peak in the Cerros de Celaque mountains rises 9,347 feet (2,849 meters) above sea level, but most of the mountains are much lower. Forests cover many slopes. Honduras has no live volcanoes, so its soils are not enriched by volcanic ash. But some of the smaller highland valleys are fertile enough to support farms.

The highlands have a milder climate than the coasts. The capital city of Tegucigalpa in the central mountains has an average temperature of 74° F (23° C). The coastal lowlands have an average temperature of 88° F (31° C).

The Northeastern Plain, sometimes called the Mosquito Coast, or Mosquitia, is Honduras's least developed and most thinly populated region. Tropical rain forests cover much of this hot, wet area. The plain has some grasslands and forests of palms and pines, as well as a few little towns.

FACTS

Official name:	Republica de Honduras (Republic of Honduras)
Capital:	Tegucigalpa
Terrain:	Mostly mountains in interior, narrow coastal plains
Area:	43,433 mi² (112,492 km²)
Climate:	Subtropical in lowlands, temperate in mountains
Main rivers:	Ulua, Chamelecón, Coco, Choluteca, Aguán
Highest elevation:	Cerro Las Minas, 9,347 ft (2,849 m)
Lowest elevation:	Caribbean Sea, sea level
Form of government:	Democratic constitutional republic
Head of state:	President
Head of government:	President
Administrative areas:	18 departamentos (departments)
Legislature:	Congreso Nacional (National Congress) with 128 members serving four-year terms
Court system:	Corte Suprema de Justicia (Supreme Court of Justice)
Armed forces:	12,000 troops
National holiday:	Independence Day - September 15 (1821)
Estimated 2010 population:	7,737,000
Population density:	178 persons per mi² (69 per km²)
Population distribution:	53% rural, 47% urban
Life expectancy in years:	Male, 68; female, 72
Doctors per 1,000 people:	0.6
Birth rate per 1,000:	27
Death rate per 1,000:	5
Infant mortality:	23 deaths per 1,000 live births
Age structure:	0-14: 38%; 15-64: 58%; 65 and over: 4%
Internet users per 100 people:	9
Internet code:	.hn
Languages spoken:	Spanish (official), Amerindian dialects
Religions:	Roman Catholic (predominant), Protestant
Currency:	Lempira
Gross domestic product (GDP) in 2008:	$13.99 billion U.S.
Real annual growth rate (2008):	4.0%
GDP per capita (2008):	$1,820 U.S.
Goods exported:	Bananas, coffee, gold, lobster, palm oil, shrimp, wood products
Goods imported:	Chemicals, food, machinery, motor vehicles, petroleum products
Trading partners:	El Salvador, Guatemala, United States

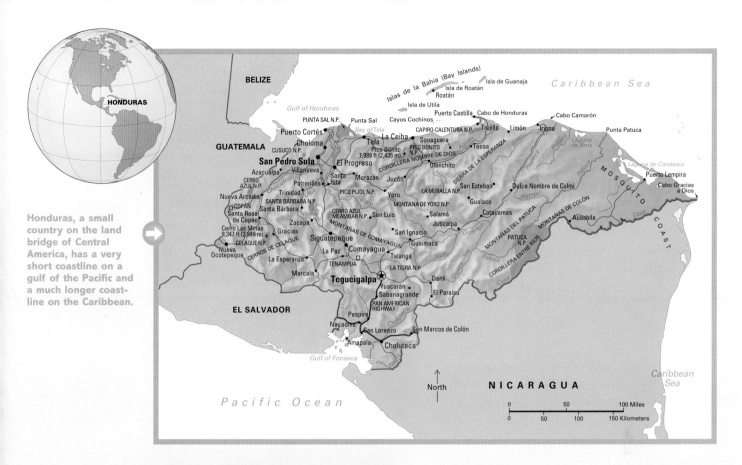

BELIZE

GUATEMALA

EL SALVADOR

NICARAGUA

Pacific Ocean

Caribbean Sea

Gulf of Honduras

Gulf of Fonseca

Caribbean Sea

Islas de la Bahía (Bay Islands)
Isla de Roatán · Isla de Guanaja
Roatán
Isla de Utila

PUNTA SAL N.P. · Punta Sal · Cayos Cochinos · Puerto Castilla · Cabo de Honduras · Cabo Camarón
Puerto Cortés · Bay of Tela · CAPIRO-CALENTURA N.P. · Trujillo · Limón · Iriona · Punta Patuca
Choloma · Tela · La Ceiba · Sonaguera · Tocoa
CUSUCO N.P. · Pico Bonito · PICO BONITO · CORDILLERA NOMBRE DE DIOS
San Pedro Sula · 7,989 ft (2,435 m) · N.P. · Olanchito · Laguna de Brus
Azacualpa · El Progreso · SIERRA DE LA ESPERANZA · Laguna de Caratasca
CERRO · Santa · Morazán · Jocón · Puerto Lempira
AZUL N.P. · Potrerillos · Rita · San Esteban · Dulce Nombre de Culmi · Cabo Gracias a Dios
Nueva Arcadia · Trinidad · PICO PIJOL N.P. · Yoro · LA MURALLA N.P. · MOSQUITO COAST
COPÁN · SANTA BÁRBARA N.P. · MONTAÑA DE YORO N.P. · Gualaco · Catacamas
Santa Rosa · Santa Bárbara · CERRO AZUL · San Luis · Salamá
de Copán · Zacapa · MEAMBAR N.P. · San Ignacio · Juticalpa · Auasbila
Cerro Las Minas · Gracias · MONTAÑAS DE COMAYAGUA · MONTAÑAS DE COLÓN
9,347 ft (2,849 m) · Siguatepeque · Guaimaca · MONTAÑAS DEL PATUCA
Nueva · CELAQUE N.P. · La Paz · Comayagua · PATUCA · CORDILLERA ENTRE RÍOS
Ocotepeque · CERROS DE CELAQUE · Talanga · N.P.
La Esperanza · TENAMPUA · LA TIGRA N.P.
Marcala · Tegucigalpa · Danlí
Yuscarán · El Paraíso
Sabanagrande
PAN AMERICAN · San Marcos de Colón
Pespire · HIGHWAY
Nacaome · San Lorenzo
Amapala · Choluteca

North

0 ———— 50 ———— 100 Miles
0 —— 50 —— 100 —— 150 Kilometers

Honduras, a small country on the land bridge of Central America, has a very short coastline on a gulf of the Pacific and a much longer coastline on the Caribbean.

Slum areas in the capital, Tegucigalpa, reveal the widespread poverty in Honduras.

The modern Legislative Palace stands in Tegucigalpa, the capital and largest city of Honduras.

The people and their government

More than 90 percent of the Honduran people are *mestizos,* people with both Indian and European ancestors. Almost all mestizos speak Spanish, and many people in the northern areas and ports also speak English.

People called Garifuna, or Black Caribs, live along the nation's northwestern coast. These people are the descendants of African slaves and Arawak Indians who lived on the Caribbean island of St. Vincent. In 1797, the British rulers of the island forced some Africans and Indians to live in Honduras because they were considered rebellious. The Garifuna speak an Arawak language, but most also speak Spanish, English, or both. About 80,000 Garifuna live in Honduras.

More than 70,000 Miskito Indians live in small communities on the Northeastern Plain. They are a mixture of native Indians, freed slaves, and other groups, and they speak the Miskito language.

Today, Honduras is an independent nation with an elected president and national legislature. Since 1981, the voters of Honduras have elected a new president and legislature every four years.

ECONOMY

Honduras is a poor country, and its people have a low average income. More than half live in rural areas, and most own or rent small farms. They live in small houses made of adobe or wood, or erected on a wooden frame packed with mud and stones.

Transportation in Honduras is limited, and communication between rural and urban areas is poor. Some modernization is taking place in cities as industry and education expand, but such changes are slow to reach the rural areas.

Agriculture

Honduras has few resources, and the nation's economy is one of the least developed in Latin America. Agriculture is one of the nation's most important economic activities. More than one-third of all Honduran workers are involved in farming.

Large numbers of Hondurans on the Northern Coast grow bananas. The banana industry was developed by U.S. companies about 1900. At that time, fruit companies cleared forests and drained swamps for banana plantations. They built railroads and ports to ship the fruit, and they established towns, hospitals, and schools for the workers.

Export taxes paid by the fruit companies to the Honduran government took care of most of the country's expenses. In return, the government gave the fruit companies special privileges. Because of this close relationship, Honduras became known as a "banana republic"—a term used to refer to countries whose economies depended on a single product.

Today, most of the banana plantations are owned by Honduran companies. However, the firms that buy the bananas and ship them to foreign markets are generally owned by foreign companies. Bananas now account for about one-tenth of the nation's exports.

Coffee, which is grown in the inland mountains, makes up about one-sixth of all Honduran exports. Other mountain crops include corn and beans. Corn, the country's basic food crop, covers more land than any other crop in Honduras. Some cattle are raised

Coconut palms in Honduras not only yield delicious fruit, but also provide building material for houses and leaves for thatching roofs and weaving baskets.

Construction of an airport extension on the Bay Islands north of the mainland occupies a tractor operator. Honduras has international airports at Tegucigalpa and San Pedro Sula.

BANANA CULTIVATION

Bananas thrive in the warm, humid parts of the world. Banana farmers start a crop by cutting growths called suckers from the underground stems of mature banana plants. The suckers are planted in the ground. In three to four weeks, tightly rolled leaves sprout up and unroll as they grow. The "trunk" of a banana plant is actually the rolled stalks of its leaves. About 10 months later, a large bud at the end of an underground stem grows from the leaf bundle. When the stem reaches the top of the plant, small flowers appear on the bud. These flowers develop into tiny green bananas. As the bananas grow, they begin to curve upward. The fruit is harvested four or five months later. Cut when still green, the bananas are packed and shipped to markets.

Processing bananas for export employs many Hondurans. Bananas are picked when green so that they will ripen during shipping to distant markets. Much of the country's income is from banana and coffee exports.

in the mountains, as well as on the plains of the Southern Coast. The fertile southern soils support cotton farms too.

Honduras exports shrimp caught near the Bay Islands off its northern coast. Timber from pines and tropical hardwood trees is also exported.

Service industries and manufacturing

More than half of Honduran workers work in service industries. Community and government services form the most important service industry group. This group includes education and health care. The trade, restaurants, and hotels industry group also employs many people. Services are mainly centered in the largest cities, especially Tegucigalpa.

Manufacturing employs only about a sixth of all the workers in Honduras. Tegucigalpa and San Pedro Sula are the major manufacturing centers. The main products are consumer goods, such as processed foods and beverages, clothing, detergents, and textiles. Pottery-making is a craft industry in rural homes. Sawmills provide lumber for furniture, paper, and wood products industries.

A small, landlocked nation in the heart of central Europe, Hungary is bordered to the north by the Czech Republic and Slovakia, to the northeast by Ukraine, to the east by Romania, to the south by Croatia and Serbia, and to the west by Austria. Before World War II (1939–1945), Hungary was an agricultural nation. When the Communists took over the government in the late 1940's, they introduced an economic plan that encouraged industrialization. Today, manufacturing and other industries contribute more to the national income than does farming.

Magyars, whose ancestors settled in what is now Hungary during the late 800's, make up most of Hungary's present-day population. The early Magyars, known as *On-Ogurs* (people of the ten arrows), were nomadic warriors trained from infancy as riders, archers, and javelin throwers. They raided the Danube and Elbe valleys and many towns throughout Europe.

The Magyars eventually converted to Christianity. In 1000, Stephen, a Roman Catholic leader of the Magyar tribe, was crowned the first king of Hungary by Pope Sylvester II, and he made Catholicism the nation's official religion. The skill of fighting on horseback lives on today in the Hungarian culture: the ancient Magyar art of fencing with a cavalry saber is still enjoyed as a national sport.

Although the country in which the Magyars settled was surrounded by Slavic and Germanic peoples, the Magyars retained their own language, which has survived to the present day as the official language of Hungary. The Magyar language—also called Hungarian—belongs to the Uralic-Altaic family of languages, which includes Estonian and Finnish. Magyar is totally unrelated to other major East European languages, such as Romanian, Polish, Czech, or Serbo-Croatian. While Magyar is spoken throughout the country, many Hungarians also speak German.

Way of life

As the country became more industrialized after World War II, the traditional life of the Hungarian peasant was replaced by more modern, urban ways. Today, more Hungarians work in industry than on farms, and about two-thirds of the Hun-

HUNGARY

garian people live in cities and towns. Budapest, the nation's capital and largest city, is home to more than 15 percent of Hungary's people.

Most Hungarians, especially city dwellers, dress in Western-style clothing, but rural people still wear colorfully embroidered costumes on special occasions. Villagers who once carved their own wooden utensils and embroidered their linens now buy manufactured household items.

Present-day Hungarians still display a traditional love of fine wines and of good, spicy food. Their most famous dish is *goulash,* a thick soup, or stew, consisting of cubes of meat, gravy, onions, and potatoes, and flavored with a seasoning called *paprika.*

A divided nation

Because the Hungarians share few linguistic or cultural links with their neighbors, they often consider themselves the most isolated people in Europe. They also regard themselves as a divided nation, since millions of Hungarians live in neighboring countries.

During World War I (1914–1918), Austria-Hungary was part of an alliance called the Central Powers. They were defeated by the Allies—a group that included the United States, the United Kingdom, France, Russia, Serbia, and other nations. Hungary and the Allies signed the Treaty of Trianon in 1920. The treaty was part of the World War I peace settlements. It stripped Hungary of more than two-thirds of its territory. Parts of Hungary went to Czechoslovakia, Romania, Austria, and the Kingdom of the Serbs, Croats, and Slovenes (later called Yugoslavia). Hungary's present boundaries are about the same as those set by the treaty.

The majority of the Hungarians living in neighboring lands are located in Romania. From the 1960's through the 1980's, Romania's Communist regime denied Hungarians fundamental human rights. The situation was a constant source of political conflict between the two countries. Today, hundreds of thousands of ethnic Hungarians also live in the Czech Republic, Slovakia, Serbia, and western Ukraine.

107

HUNGARY TODAY

October 23, 1989, saw the birth of the new Republic of Hungary—a parliamentary democracy based on a free market economy. The nation's gradual movement toward democracy began with the economic reforms of János Kádár in the 1960's. Kádár was head of Hungary's Communist Party and served as premier from 1956 to 1958 and from 1961 to 1965. He tried to win the support of the people by easing some of the restrictions on cultural, economic, and social life. In 1968, the government introduced a new economic program that combined features of a free market system with the country's socialized economy.

These policies came to be known as *goulash Communism* because of the way elements from different systems were combined—just like the Hungarian national dish. Kádár's reorganization soon began to improve the economy, and living standards rose. However, as a result of the new economic freedoms, some people became considerably more wealthy than others, and the prices of many popular goods rose beyond the reach of the poorest people.

In addition to the economic reforms, Hungarians also enjoyed more personal freedom. They were allowed to travel to Western countries if they could obtain the necessary foreign currencies—a major drawback for most Hungarians. Cultural life was also relatively free from government restrictions during this period, and Hungarians were allowed to express their views without fear of censorship or punishment.

During the 1970's, however, the Hungarian economy worsened, and living standards began to decline. In May 1988, Kádár was forced to step down, and Károly Grósz replaced him as head of the Communist Party.

FACTS

Official name:	Magyar Koztarsasag (Republic of Hungary)
Capital:	Budapest
Terrain:	Mostly flat to rolling plains; hills and low mountains on the Slovakian border
Area:	35,919 mi^2 (93,030 km^2)
Climate:	Temperate; cold, cloudy, humid winters; warm summers
Main rivers:	Danube, Tisza
Highest elevation:	Mount Kékes, 3,330 ft (1,015 m)
Lowest elevation:	Near Szeged, 259 ft (79 m)
Form of government:	Parliamentary democracy
Head of state:	President
Head of government:	Prime minister
Administrative areas:	19 megyek (counties), 6 cities that rank as counties
Legislature:	Orszaggyules (National Assembly) with 386 members serving four-year terms
Court system:	Supreme Court
Armed forces:	25,200 troops
National holiday:	Saint Stephen's Day - August 20
Estimated 2010 population:	9,970,000
Population density:	278 persons per mi^2 (107 per km^2)
Population distribution:	67% urban, 33% rural
Life expectancy in years:	Male, 69; female, 77
Doctors per 1,000 people:	3.0
Birth rate per 1,000:	10
Death rate per 1,000:	13
Infant mortality:	6 deaths per 1,000 live births
Age structure:	0-14: 15%; 15-64: 69%; 65 and over: 16%
Internet users per 100 people:	55
Internet code:	.hu
Languages spoken:	Magyar (also called Hungarian) is the official language
Religions:	Roman Catholic 51.9%, Calvinist 15.9%, Lutheran 3%, Greek Catholic 2.6%, Jewish 1%, other 25.6%
Currency:	Forint
Gross domestic product (GDP) in 2008:	$154.67 billion U.S.
Real annual growth rate (2008):	-1.5%
GDP per capita (2008):	$15,436 U.S.
Goods exported:	Aluminum, electronic equipment, machinery, petroleum products, pharmaceuticals, steel, transportation equipment
Goods imported:	Chemicals, electric power, food, machinery, petroleum products, transportation equipment
Trading partners:	Austria, France, Germany, Italy, Russia

Hungary, a land-locked nation in central Europe, fell under Communist control after World War II. A people's revolt in 1956 was quickly crushed by Soviet troops. On the 33rd anniversary of the 1956 uprising, Hungary declared itself a parliamentary democracy.

A ferryboat on the Danube River passes by the domed Parliament Building, one of Budapest's historic landmarks. Hungary enjoys a thriving tourist trade.

In January 1989, the National Assembly, Hungary's legislature, passed a law guaranteeing Hungarians the right to demonstrate freely. Another new law allowed Hungarians to form associations and political parties independent of the Hungarian Socialist Worker's Party (HSWP), the Communist Party in Hungary.

Hungary's first multiparty elections were held in 1990. They resulted in a bitter defeat for the Communists and a victory for the Democratic Forum. Arpád Göncz became president, and József Antall became prime minister. The president is the head of state and the country's most powerful government official. The prime minister serves as the head of government.

Goncz was reelected president in 1995. He was succeeded by Ferenc Mádl in 2000 and by László Sólyom in 2005. A former Olympic fencing gold medalist, Pál Schmitt, was elected president in 2010. In 2011, Schmitt signed a new constitution to replace the 1949 charter. Prime Minister Viktor Orbán claimed the new constitution would complete Hungary's transition from Communism to democracy. But human rights groups criticized the new document, saying it failed to protect citizens' rights.

HISTORY

The history of the Hungarian state began in the late 800's with the arrival of the Magyars, who settled in the middle Danube Basin. The Danube Basin is the great lowland region bordering the Danube River that makes up most of present-day Hungary. The Magyar tribes were led by Árpád, whose great-grandson, Géza, organized the tribes into a united nation about 100 years later.

When Géza's son Stephen became the first king of Hungary, he brought his Roman Catholic faith to the country. During Stephen's reign, Hungary became closely identified with the culture and politics of Western Europe. When the last Árpád king died in 1301, Hungary was firmly established as a Christian state and remained an independent kingdom for another 225 years.

The Hungarian Empire

Hungary reached the height of its political power and cultural influence under Matthias Corvinus Hunyadi, who ruled from 1458 to 1490. During that time, the country became a center of the Renaissance, the great artistic and cultural movement that spread across Europe during the 1400's and 1500's.

A period of conflict and disorder followed the death of Matthias Corvinus Hunyadi in 1490, and the Hungarian Empire became seriously weakened. In 1526, the Ottomans seized control of the eastern third of the country, called Transylvania, and made it a principality dependent on them. The Austrian Habsburgs took the country's western and northern sections. Then, in the late 1600's, the Habsburgs drove the Ottomans out of Hungary and seized control of the entire country.

The harsh rule of the Habsburgs led to a nationwide revolt in 1703, led by Francis Rákóczi II, whose family included princes of Transylvania. The Habsburgs had crushed the rebellion by 1711, but they were persuaded to relax their rule and improve conditions in Hungary.

During the 1840's, revolutions broke out across Europe, and Hungary tried to break away from Austria. Led by Louis Kossuth, the Hungarians proclaimed their independ-

Late 800's	The Magyars conquer Hungary.
1000	Stephen I becomes Hungary's first king and converts the country to Roman Catholicism.
1241	The Mongols invade Hungary.
1458-1490	Matthias Corvinus Hunyadi rules Hungary, which becomes a center of Renaissance culture.
1514	Hungarian nobles crush a peasant revolt.
1526	The Ottomans defeat Hungary in the Battle of Mohács and soon occupy central and eastern Hungary.
1600's	Austrian Habsburg forces drive the Ottomans out and take control of Hungary.
1703-1711	Francis Rákóczi II leads an unsuccessful uprising for independence.
1848	Louis Kossuth leads an anti-Habsburg revolution, which is defeated the following year.
1867	The Dual Monarchy of Austria-Hungary is established.
1914-1918	Austria-Hungary is defeated in World War I.
1918	Hungary becomes a republic.
1919	Béla Kun establishes the first Hungarian Communist government, which lasts only a few months.
1919-1944	Admiral Nicholas Horthy rules Hungary as regent.
1920	Under the Treaty of Trianon, Hungary loses two-thirds of its territory.
1941	Hungary enters World War II on Germany's side.
1944	Germany occupies Hungary.
1945	Hungary and the Allies sign an armistice.
1946	Hungary becomes a republic, and the new government introduces political, economic, and social reforms.
1946-1949	Hungarian Communists gradually gain control of the government.
1947	The Allies sign a peace treaty with Hungary that confirms the terms of the 1945 armistice.
1955	Hungary becomes a member of the United Nations.
1956	Soviet forces crush an anti-Communist revolution in Hungary.
198	Hungary's Communist Party agrees to allow other political parties to operate in the country.
1989	Hungarians gain the right to demonstrate freely.
1990	Democratic elections are held.
1999	Hungary becomes a member of the North Atlantic Treaty Organization.
2004	Hungary becomes a member of the European Union.

Stephen I, first King
(975?-1038)

Franz Liszt, composer
pianist
(1811-1886)

Béla Bartók, composer
(1881-1945)

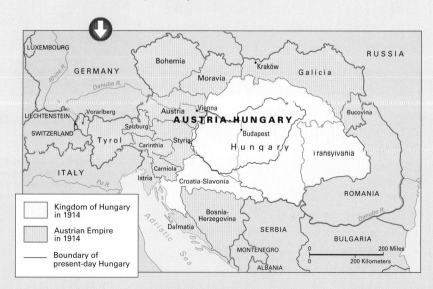

Armed citizens patrol Budapest during the anti-Communist uprising of October 1956. A month later, Soviet forces invaded Hungary and brutally crushed the revolt. Many Hungarians were killed or imprisoned.

ence. The revolution was eventually crushed by the Habsburgs, but only with the help of Russian troops. Once again, Hungary found itself under Austrian rule.

As Austrian power declined during the 1800's, the Hungarians, led by Francis Deák, were able to force the emperor of Austria to establish the *Dual Monarchy*. Under this arrangement, both Austria and Hungary had the same monarch and conducted foreign, military, and certain financial affairs jointly. But each country had its own government to handle all other matters.

The creation of the Dual Monarchy of Austria-Hungary in 1867 led to great prosperity in Hungary. Over the next 50 years, the nation's economy, educational system, and cultural life developed rapidly.

The end of Austria-Hungary

In 1918, Austria-Hungary's defeat in World War I brought an end to the Dual Monarchy. Thirteen days after Austria-Hungary signed an armistice with the Allied powers, the Hungarian people revolted, declaring Hungary a republic. After a brief period under a Communist dictatorship, Hungary again became a monarchy under Admiral Nicholas Horthy. He ruled as *regent* (in place of a monarch) from 1919 until 1944, when Hitler's troops seized the country and set up a Nazi government.

The Soviet Union invaded Hungary late in 1944, and Hungary and the Allies signed an armistice in January 1945. Early in 1946, Hungary was declared a republic, but the Communists gradually gained control of the government. By 1947, Matthias Rákosi, head of the Communist Party, ruled as dictator. Rákosi's policies resulted in a period of severe persecution for the Hungarian people that did not end until the 1960's.

In the late 1980's, the reform movement in Hungary gained strength. Public pressure forced the Communist Party's leaders to allow other political parties to form. In 1990, Hungary held its first multiparty elections since 1949.

A magnificent crown now preserved in the National Museum of Budapest is said to be that of King Stephen I, who was made a saint in 1083 for bringing Roman Catholicism to Hungary.

Hungary once stood at the center of the empire of Austria-Hungary--the Dual Monarchy that reached its greatest height before World War I. After the heir to the Austro-Hungarian throne was assassinated in 1914, Austria-Hungary declared war on Serbia, touching off World War I.

ENVIRONMENT

Most of Hungary's land is low. Eastern Hungary is almost entirely flat, while western Hungary consists mainly of rolling hills and low mountains. The Tisza—the country's longest river and a tributary of Hungary's most important river, the Danube—flows from north to south through eastern Hungary.

Land regions

Hungary's four main land regions are the Great Plain, Transdanubia, the Little Plain, and the Northern Highlands. The Great Plain stretches across all of Hungary east of the Danube, except for the mountains in the north. This flat plain, broken only by river valleys, sand dunes, and small hills, covers about half of Hungary's total land area.

In the 1500's and 1600's, when Hungary fell under Ottoman rule, the fertile farmland of the Great Plain was a *puszta* (steppeland), where cattle grazed and wild horses roamed. Today, parts of the puszta are preserved at Hortobágy, a national park about 100 miles (160 kilometers) east of Budapest.

Crops of sugar beets, melons, sunflowers, and wheat grow on the Great Plain. With modern irrigation, orchards and vineyards now flourish in the sandy areas. And swamps that once provided a habitat for waterfowl have been drained to provide farmland for corn production.

Transdanubia, a region that consists mostly of hills and mountains, covers all of Hungary west of the Danube, except for the northwest corner of the country. The Transdanubian Central Highlands—a chain of low, rounded mountains that includes the Bakony and Vertese ranges—extend along the entire northern side of Lake Balaton.

A region of great scenic beauty, the Transdanubian Central Highlands feature mountain streams, great oak forests, and picturesque ravines plunging between chalk and dolomite cliffs. The foothills of the Austrian Alps rise in the western region of Transdanubia, while the southeastern part is a major farming region.

The smallest region, the Little Plain, occupies the northwest corner of Hungary. The land in this area is flat, except for the foothills of the Austrian Alps along the western boundary. The steep slopes of the Northern Highlands rise northeast

Farmworkers use long-handled hoes to clear the ground of weeds. When the Communists controlled Hungary, most farming was done on state farms owned by the government and on collective farms where many families worked together.

A worker in Gyoer, west of Budapest, assembles a diesel engine for installation in an automobile. Transportation equipment ranks among Hungary's leading exports.

A man in traditional costume drives a team from the backs of two of the horses. Roman, or trick, riding is one of the tourist draws of Hungary's puszta conservation area at Hortobágy.

A vendor sells produce at a food market in Budapest. Hungary's chief crops include apples, barley, corn, potatoes, sugar beets, sunflowers, wheat, and grapes for wine.

Often called the Hungarian Sea, Lake Balaton is central Europe's largest lake, covering about 230 square miles (596 square kilometers). Averaging 10 feet (3 meters) deep, the lake is easily warmed by the sun, making it a popular recreation spot.

of the Danube and north of the Great Plain. This region is densely forested, with small streams and dramatic rock formations.

Agriculture and industrialization

Hungary's most important natural resources are fertile soil and a climate that is generally favorable for agriculture. Although the nation has been heavily industrialized since World War II, farming remains an important industry. In addition to growing crops such as apples, barley, corn, potatoes, sugar beets, sunflowers, wheat, and wine grapes, Hungary's farmers raise cattle, chickens, hogs, and other livestock.

Factories in Hungary produce iron and steel, buses and railroad equipment, electrical and electronic goods, food products, pharmaceuticals, medical and scientific equipment, and textiles. Hungary is also one of the world's leading producers of bauxite.

In the late 1980's, Hungary began the process of changing from a socialized economy to a free market economy. The government encouraged private industry, and foreign investors began to bring businesses and jobs to the country.

In 2004, Hungary joined the European Union, an organization of European countries that promotes economic and political cooperation among its members. However, the nation continues to face many economic challenges.

BUDAPEST

No other city in Hungary can match the splendor of Budapest. The capital and largest city, it is also the center of Hungary's culture and industry. It represents the spiritual and cultural home of the millions of Hungarians living outside the country.

A union of cities

Situated on both banks of the Danube River in northern Hungary, modern Budapest consists of the once adjoining cities of Buda, Pest, and Óbuda. The city also includes Margaret Island, in the Danube River.

Budapest covers 203 square miles (525 square kilometers). Eight bridges connect the eastern and western banks of the Danube in Budapest.

The part of the city that used to be the city of Buda rests high on the west bank of the river. Buda was the permanent residence of the Hungarian kings beginning in the 1300's. Standing on steep, wooded hills and crowned by Castle Hill, Buda reflects Hungary's ancient military past. The Royal Palace, which includes the remains of an ancient fort, stands on top of Castle Hill.

In contrast, Pest, on the east side of the river, represents Hungary's coming of age as a modern European state. Most of Pest was developed in the 1800's on a series of plateaus. Its public buildings and boulevards rival those of Paris and Vienna in their style and elegance. The Hungarian Academy of Sciences and the House of Parliament are located in the old Inner City, called Belváros.

Budapest was united in 1873, when Hungary was part of the Dual Monarchy of Austria-Hungary. At that time, the city was the scene of rapid developments in mechanical engineering, as well as in the milling and iron industries. Shipyards, breweries, and tobacco-processing factories also prospered. As agricultural and foreign workers flocked to find jobs in the city, Budapest's population increased from about 178,000 in 1850 to about 1,100,000 in 1910.

The golden age

The era of the Dual Monarchy is often regarded as Budapest's architectural golden age. The Hungarians wanted Budapest to be as splendid as Vienna, the capital of the Habsburg Empire, and Budapest took on its present-day character during this period.

The old heart of the city in Pest was torn down because the Baroque and neoclassical buildings erected between the 1500's and 1700's were considered too small and old-fashioned. They were replaced by buildings whose architecture was influenced by the grandest styles of earlier times. For example, the domed Parliament Building was designed in the neo-Gothic style and based on medieval models.

Budapest is situated on both banks of the Danube River. A series of bridges connects the city's eastern and western sections.

The relaxing thermal waters of the Gellért Baths, built between 1912 and 1918 in the Art Nouveau style, are one of Budapest's famous tourist attractions.

The ornate splendor of a Budapest restaurant dates from the late 1800's, when the city was an important center of Austria-Hungary.

The Chain Bridge is one of Budapest's most famous landmarks. The magnificent suspension bridge, which opened in 1849, connected Pest and Buda, at the time still separate cities.

The Royal Palace is neo-Baroque, while other buildings followed the Italian-inspired styles of the 1400's and 1500's. Budapest also boasts several houses and hotels in the Art Nouveau style of the late 1800's and early 1900's, which have been greatly admired for their beauty and brilliant colors.

In 1887, the first electric trams appeared in Budapest, followed only nine years later by an underground railway beneath the city's main boulevard. Residents of Budapest were the first Europeans to travel underground. Today, old-style yellow carriages still clatter along the line. However, the modern Metro, which runs under the Danube between Buda and Pest, is quite different. Modeled after the impressive Moscow Metro, it serves as one of the main arteries in the city's transport system.

Today, Budapest has an old-world, turn-of-the-century charm. Budapest's distinguished past lingers on in the city's coffee houses, with their ornate ceilings, gold columns, and marble tables. The citizens of Budapest enjoy meeting in these old coffee houses, where they read or chat with friends.

EAST EUROPEAN ROMA

Roma are a nomadic people whose ancestors originally lived in India. They are sometimes called Gypsies, Romanies, or Travellers. About A.D. 1000, they left India and wandered westward through the Middle East, arriving in Europe at the beginning of the 1300's.

Some Roma claimed to have come from a country called Little Egypt. The word *Gypsy* is a shortened form of *Egyptian*. However, most groups prefer the name *Roma* for the people in general, because *Gypsy* has sometimes been used as an insult. Today, Roma live in all parts of the world, but the largest numbers are found in eastern Europe. According to some estimates, there are about 12 million Roma living throughout the world.

Because Roma have traditionally been a wandering people who chose to live outside the mainstream of society, little is known of their history and culture. As a result, there is an air of romantic mystery about them, and their colorful costumes, lively music, and dancing skills have fascinated people for generations. At the same time, however, they have suffered a great deal of persecution. Even today, Roma experience high rates of poverty, unemployment, and illiteracy.

A history of persecution

At first, the Roma were welcomed in Europe. The European nobility admired—and exploited—their iron-working skills. The Spanish rulers Ferdinand and Isabella may have used Roma-made weapons to defeat the Moors at Granada in 1492. In Hungary, Roma were employed in the manufacture of instruments of torture as well as the making of weapons.

Although the Spanish Roma were poorly rewarded for their work, they fared much better than the Hungarian Roma, who were slaves to the Magyar kings. The Roma were enslaved by the Romanian nobility, who needed laborers to work their vast estates, and by the Romanian clergy, who believed that Jesus Christ had cursed the Roma. Sold at auction in slave markets, the Roma were forced to work under the most brutal conditions.

As the years went by, prejudice against the Roma grew, and they were blamed for a variety of crimes, from theft to kidnapping. In 1782, Hungarian Roma were even accused of cannibalism, and many were driven to the swamps and drowned by Hungarian soldiers.

During World War II (1939-1945), Adolf Hitler condemned the Roma, along with various other religious and ethnic groups, as "racially impure." The Nazis rounded up the Roma and imprisoned them in concentration camps. Hundreds of thousands of Roma were murdered in these camps during the war. After the Communists gained control of eastern Europe, they, too, condemned the Roma for their failure to be contributing members of a socialist society.

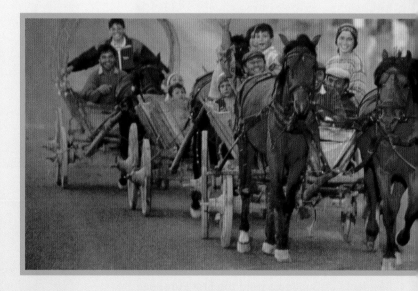

Romani violinists entertain the crowds at a country fair in Hungary. The violin has a special place in Romani music, and the Roma believe that the wood for the first Romani violin came from the dense forests of Transylvania.

A love of music and dance

Roma have served as entertainers since the late 1400's, when they were employed as musicians at the court of Matthias Corvinus Hunyadi. Touring bands of Roma with trained bears once provided entertainment in remote Romanian villages. The Magyar language even has a special word for being entertained by Roma. It is *cigányozni (Gypsying)*.

Roma played not only at banquets and special occasions, but also at military events. In the Austro-Hungarian army, recruiting officers had Romani musicians play tunes called *verbunkos* to stir their patriotic spirit, and Roma also provided the music that led troops into battle.

The Roma of southern Spain were the first to perform the *flamenco* style of dance and music that is still popular today. Flamenco dancing, accompanied by guitars and castanets, may include much skillful footwork, finger snapping, and forceful but flowing arm movements.

Romani music played an important role in the works of Franz Liszt, the Hungarian pianist and composer. His *Hungarian Rhapsodies,* in which he used Romani music and folk-dance themes, are among his best-loved compositions. In describing Romani music, Liszt wrote, "The chief characteristic of this music is the freedom, richness, variety, and versatility of its rhythms, found nowhere else in a like degree."

A tame brown bear stands in the glow of a Romani campfire. Trained bears were once part of the Roma's traveling shows. In addition to providing entertainment, many Roma have made their living as fortunetellers and horse traders.

This Romani family in Tulcea, Romania, travels by horse-drawn wagon. Many Romani families consist of a husband and wife, their unmarried children, their married sons, and the sons' wives and children.

Roma in colorful costumes, like this woman in bright orange, take part in a festival in Hungary.

A Romanian Romani woman wears the flowery headscarf characteristic of the region. Most Roma speak the language of the people among whom they live, but many also speak their own native tongue, often called Romani.

ICELAND

An island country in the Arctic and North Atlantic oceans, Iceland lies just south of the Arctic Circle between Greenland and Norway. Many people call Iceland the *Land of Ice and Fire* because of its unusual landscape, where large glaciers lie next to steaming hot springs, geysers, and volcanoes. With only 8 persons per square mile (3 persons per square kilometer), this remote island is the most sparsely populated country in Europe.

Iceland's first settlers arrived about 870, with the Norwegian adventurer Ingólfur Arnason. Later, more settlers came from Norway and from Viking colonies in Britain and Ireland. Present-day Icelanders resemble the people of Ireland, Scotland, and the Scandinavian countries of Denmark, Norway, and Sweden. Nearly all of Iceland's people live close to the coast, which is warmed by the Gulf Stream.

About a seventh of Iceland's working people catch fish for a living or are employed in fish-processing plants. Icelanders enjoy a high standard of living. Much of what they buy must be imported, so the cost of living is quite high. As a result, most women work outside the home, and many Icelanders hold more than one job.

Despite their modern lifestyle, Icelanders have kept many connections to the past. Their Icelandic language is still so similar to the Old Norse language spoken by their ancestors that people today can easily read tales and poems written in the 1100's and 1200's. In the tradition of their ancestors, Icelanders do not have family names. Instead, they take on a second name by simply adding -*son* (son) or -*dóttir* (daughter) to their father's first name. Women do not change their names when they marry.

Today's Icelanders share a long, eventful history troubled by natural disasters and political unrest. In the 1200's, civil wars between the early settlers ended only when the *Althing* (parliament) agreed to accept the king of Norway as the ruler of Iceland.

FACTS

Official name:	Lyoveldio Island (Republic of Iceland)
Capital:	Reykjavík
Terrain:	Mostly plateau interspersed with mountain peaks, icefields; coast deeply indented by bays and fiords
Area:	39,769 mi² (103,000 km²)
Climate:	Temperate; moderated by North Atlantic Current; mild, windy winters; damp, cool summers
Main rivers:	Jökulsa à Fjöllum, Hvitá, Skálfandafljót, Thjórsá
Highest elevation:	Hvannadalshnúkur, 6,952 ft (2,119 m)
Lowest elevation:	Atlantic Ocean, sea level
Form of government:	Republic
Head of state:	President
Head of government:	Prime minister
Administrative areas:	23 syslar (counties), 14 kaupstadhir (independent towns)
Legislature:	Althing (Parliament) with 63 members serving four-year terms
Court system:	Haestirettur (Supreme Court)
Armed forces:	None
National holiday:	Independence Day - June 17 (1944)
Estimated 2010 population:	314,000
Population density:	8 persons per mi² (3 per km²)
Population distribution:	92% urban, 8% rural
Life expectancy in years:	Male, 79; female, 83
Doctors per 1,000 people:	3.8
Birth rate per 1,000:	14
Death rate per 1,000:	7
Infant mortality:	2 deaths per 1,000 live births
Age structure:	0-14: 21%; 15-64: 67%; 65 and over: 12%
Internet users per 100 people:	66
Internet code:	.is
Languages spoken:	Icelandic, English, Nordic languages, German
Religions:	Lutheran Church of Iceland 80.7%, Roman Catholic Church 2.5%, Reykjavik Free Church 2.4%, Hafnarfjorour Free Church 1.6%, other 12.8%
Currency:	Icelandic Krona
Gross domestic product (GDP) in 2008:	$17.74 billion U.S.
Real annual growth rate (2008):	-3.5%
GDP per capita (2008):	$59,141 U.S.
Goods exported:	Aluminum, animal products, fish and fish products
Goods imported:	Chemicals, food, machinery, motor vehicles, petroleum and petroleum products
Trading partners:	Denmark, Germany, Netherlands, United Kingdom, United States

The Central Bank building stands in Reykjavik, Iceland's capital and largest city. The bank includes a library, an archive, and a coin collection.

Reykjavík is located on the southwest coast of the island. The city's buildings are heated by water piped in from nearby hot springs by the flow of gravity.

Iceland officially gained its independence from Denmark on June 17, 1944. The proclamation was signed at the historic site of Thingvellir, where the Althing—the world's first parliament—was established in 930.

When Norway united with Denmark in 1380, Iceland came under Danish rule. From 1402 to 1404, the *Black Death* (bubonic plague) swept across the island, killing about a third of the population. In the late 1700's, volcanic eruptions destroyed livestock and huge areas of farmland, causing massive starvation. During the Napoleonic Wars of the early 1800's, ships bringing food could not reach Iceland, and many people starved.

Iceland has been a republic since gaining its independence in 1944. During the late 1980's, a reduced fish catch and a worldwide ban on whale hunting damaged the nation's economy. To reduce its dependence on imports, Iceland expanded its range of manufactured products. Service industries, such as banking and publishing, also expanded.

In 2008, worsening world economic conditions strained Iceland's financial sector, which had become central to the nation's economy. Three large banks collapsed, some small businesses failed, and unemployment rose. In the wake of the crisis, Jóhanna Sigurdardóttir was appointed as Iceland's first female prime minister in 2009.

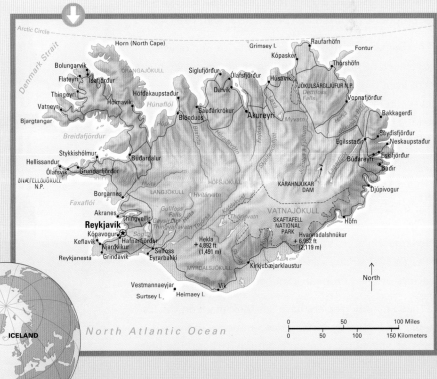

ENVIRONMENT

Iceland's natural wonders range from steaming geysers and hot springs to glistening lava fields and fiery volcanoes. This unusual landscape has been shaped by the fault line that runs across the island.

The fault line represents a boundary between two gigantic plates in Earth's crust—the North American Plate and the Eurasian Plate. The gradual movement of the plates along the fault line produces a strain on the rocks on either side of the fault. The strain of the motion often causes earthquakes.

Iceland was formed millions of years ago, when undersea lava flowing from a ridge on the seabed reached the surface of the sea. As recently as 1963, a volcano erupted south of Iceland, forming a new island called Surtsey. Iceland itself has about 200 volcanoes, and their eruptions have spread lava and volcanic rock over the land throughout the island's history.

Volcanoes and geysers

Iceland's most famous volcano is Mount Hekla, which rises 4,892 feet (1,491 meters) above sea level. The main crater of the volcano had been quiet for more than 100 years when it suddenly erupted in 1947. The eruption lasted for 13 months, spreading fiery lava over an area of 15 square miles (39 square kilometers). Similar eruptions occurred in 1980, 1981, and 1991.

Accompanying this volcanic activity are the geysers—explosions of hot water, cloudy with steam, which burst forth from the ground in huge columns. The word *geyser*, in fact, comes from the name of Iceland's most famous hot spring, *Geysir*, which spouts water about 195 feet (59 meters) into the air. Only 70 miles (110 kilometers) from the capital city of Reykjavík, dozens of geysers appear within a circle of 10 miles (16 kilometers).

Glaciers cover about one-eighth of Iceland's surface. The largest glacier, Vatnajökull, covers 3,130 square miles (8,100 square kilometers) and is as big as all the glaciers on the European continent combined.

These mighty glaciers have left their mark on the Icelandic landscape by carving deep trenches in the bottoms of many fiords, creating natural harbors. The holes

High pastures fringe an icefield on the edge of Iceland's inland plateau. The lava flow in the foreground is a reminder of the island's volcanic origins. Iceland has about 200 active volcanoes, and there are also active volcanoes under the sea off the coasts.

After lying dormant for more than 5,000 years, a volcano on the island of Heimaey off Iceland's southern coast, erupted in bursts of fiery lava in 1973. The volcano poured ash over Heimaey's only town, Vestmannaeyjar, forcing the evacuation of all the residents. The lava flow enlarged the island by almost 1 square mile (2.6 square kilometers), and after the lava had cooled and solidified, it was found to have improved the shape of the island's main harbor. The harbor's entrance was narrowed, thus providing greater protection to boats within the harbor.

made by glaciers on land have made numerous lakes, while water from melting glaciers forms rushing rivers and spectacular waterfalls. The most beautiful waterfalls in Iceland are Gullfoss in the south and Dettifoss in the north.

Climate and land regions

The warm Gulf Stream current reaches Iceland from the south, warming the coastal lowlands throughout the year and keeping the harbors free of ice. As a result, Iceland enjoys a comparatively mild climate, with frequent rainfall, patchy sunlight, and medium-force winds. Summers are mild and winters are cool.

Iceland's many volcanoes have formed a huge inland plateau, which covers most of the interior of the island. Here, the land surface is made up of *basalt* (hard volcanic rock), and lumps of volcanic lava create a barren, almost unearthly, landscape. Little vegetation can take root in this region because the porous volcanic surface discourages plant growth.

The coastal areas, by contrast, contain fertile farmland. The chief crop is hay, which is used to feed sheep, cattle, and the small Icelandic horses. Heavy rainfall and the long hours of summer sunshine allow Icelandic farmers to grow two or three crops of hay each year. Some farmers also grow root crops, such as turnips and potatoes. The nation has more than 4,000 farms.

Námafjall is a geothermal area in a volcanic area in northern Iceland. Námafjall features boiling mudpools and fumaroles, which are holes in the Earth's crust that emit steam and hot gases.

In southwestern Iceland, the Svartsengi geothermal power station taps energy from deep within Earth's crust. Many people bathe in the warm blue waters nearby, believing that dissolved minerals help treat skin conditions and breathing problems.

Fish are processed in a factory near Reykjavík. Fishing crews in large trawlers (fishing boats) drag nets along the seabed to catch capelin, cod, haddock, and herring.

ICELANDIC SAGAS

According to Norse mythology, two places existed before the creation of life—Muspellsheim, a land of fire, and Niflheim, a land of ice and mist. Between them lay Ginnungagap, a great emptiness where heat and ice met. Where Muspellsheim and Niflheim met, fire thawed ice to set the stage for the creation of the world.

The Viking seafarers who discovered Iceland must have thought the old legend had come to life when they saw the island's strange landscape. Towering glaciers and bleak stone deserts, like those of Niflheim, lay alongside active volcanoes and fiery lava, like those of Muspellsheim.

Ancient Viking tales

These Norse adventurers brought with them a wealth of Scandinavian tradition, including poems and stories that were told and retold around roaring fires during the long, dark nights of winter. Gradually, these age-old tales were carefully written down, then copied and recopied through the years.

These stories and poems gave rise to the *sagas*, a great body of literature written in Iceland between the 1100's and the 1300's. The word *saga* is related to the Icelandic verb *to say* or *to tell*. Beginning in the 1100's, Icelanders such as Saemund the Learned, Ari Thorgilsson, and Snorri Sturluson set down in writing the tales of their Scandinavian ancestors, as well as detailed records of their own Icelandic history.

The great poet and historian Snorri Sturluson wrote *Heimskringla (Circle of the World)*—a history of the kings of Norway from their origins up to his own day. He probably also wrote *The Saga of Egill Skallagrimsson*, one of the best Icelandic sagas, about a great poet of the 900's who was one of Snorri's forefathers.

Snorri also composed the *Prose Edda,* whose first section narrates myths about Scandinavian gods. Two centuries after Iceland officially converted to Christianity, Snorri Sturluson gave his people an extraordinary collection of the beliefs of his ancestors, leaving a dramatic picture of Viking gods and heroes for generations to come.

A cloud of steam drifting across the barren volcanic landscape of Iceland—which is virtually unchanged since the sagas were written—brings to mind the Norse creation myth, which speaks of a place where the land of frost meets the land of fire.

Fish are dried on a large framework before being preserved in salt. The Norse seafarers, whose adventures live on through the sagas, depended heavily on preserved fish during their long voyages.

The Vikings were fierce pirates and daring sailors who explored the North Atlantic Ocean from the late 700's to about 1100. Vikings were among the best shipbuilders of their time. They constructed vessels for trade and warfare from wood cut from the plentiful Scandinavian forests. Their trading ships, called knorrs, were about 50 feet (15 meters) long.

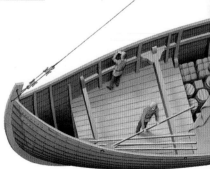

The classic stories

The classic stories, known in English as *Icelandic Family Sagas,* were composed in the 1200's by anonymous writers. These sagas vary in length from brief stories to the equivalent of full-length novels. They relate the adventures of Icelandic and Scandinavian heroes—famous people of the past who had a passion for honor, a respect for poetry and learning, and a fierce desire for independence.

The sagas bring to vivid life the people who gave Iceland its dramatic and eventful past. Icelanders still enjoy the tale of Helgi the Lean, who "believed in Christ, but prayed to Thor [a Norse god] for sea voyages and in times of danger." And people today are still inspired by the story of Njal Thorgeirsson, who strove for peace in a violent world, achieving it only by sacrificing himself and his family.

Scholars once thought that the classic sagas had been told and retold for generations before being written down during the 1200's. Today, however, most experts believe the sagas were artistic creations based on oral and written traditions. They were written during a period of moral and social decline, and they upheld the high values of a previous "golden age" that occurred between 850 and 1050.

India, the seventh largest country in the world in area, is a huge peninsula, extending far out into the Indian Ocean. High mountain ranges extend along India's northern border and separate most of the country from the rest of Asia. Pakistan lies on India's northwest border. To the northeast, India borders China, Nepal, Bhutan, Myanmar, and Bangladesh.

With more than a billion people, India is the world's second largest country in population. Only China has more people. The river valleys of northern India are among the most densely populated places in the world.

For centuries, the Western world looked upon India as a land of mystery, wealth, and excitement. The glories of its past can be seen in the Taj Mahal, at Agra in northern India. This splendid white marble structure was built by a Mughal ruler between about 1630 and 1650 as a tomb for his favorite wife. About 20,000 workers built the structure, which, according to tradition, was designed by a Turkish architect. It is still one of the world's most famous and beautiful buildings.

Today, India is a fascinating mixture of old and new. Traffic on busy city streets is often stopped by wandering cattle, which Hindus consider sacred and allow to roam freely. Many Indians go to work in modern factories wearing traditional clothing and carrying supplies on their heads in handwoven baskets.

India is a scenic land of dramatic contrasts and variety. Along its northern border lies the Himalaya, the tallest mountain system in the world. Jungles on the southern peninsula are the natural habitat of elephants and other large animals. The northern plains include the world's largest *alluvial plain* (land formed of soil left by rivers).

The people of India belong to many different ethnic groups, and their way of life varies widely from region to region. They speak about 18 major languages and a total of over 1,000 languages and dialects.

Some Indians are very wealthy, but many others are so poor that they have no homes and must sleep in the streets. Some Indians are college graduates, and India has over 8,000 universities and colleges. Yet there are also many children who must leave school at an early age to help on the family farm or to get a job.

INDIA TODAY

A British colony since the late 1700's, India won its independence on Aug. 15, 1947. After independence, India's government worked to improve the country's standard of living. Agricultural and industrial production increased. More children went to school. New laws expanded the rights of the country's most oppressed people.

Yet many problems still face India. Although the country has many well-educated and prosperous people, others are still desperately poor. Regional, language, and ethnic differences among Indians have created many obstacles for national unity.

During India's struggle for freedom in the early 1900's, some Hindus and Muslims began to fight each other to gain political power for their own group. In an effort to stop the fighting, Indian and British leaders divided the country into two nations: India, for the Hindus, and Pakistan, for the Muslims. Both new countries claimed Kashmir, an area on the northern border of India. Kashmir's ruler was Hindu, but most of its people were Muslims. Kashmir has been the focus of the struggle between Muslims and Hindus since 1947. When Pakistan invaded the area, Kashmir's ruler made it part of India for protection. War raged on until 1949, when the United Nations arranged a cease-fire.

In the 1980's, followers of the Sikh religion demanded more political control of their home state of Punjab. In 1984, India's prime minister Indira Gandhi was assassinated by two Sikh members of her security force. After her assassination, the Congress-I party chose her son Rajiv Gandhi as prime minister. Rajiv himself was assassinated in 1991.

In the 1980's and 1990's, a number of ethnic separatist groups emerged, including those in Assam and Manipur. In the state of Jammu and Kashmir, war broke out between security forces and Muslim guerrillas seeking independence.

Violent incidents between Hindus and Muslims also occurred in other parts of India during the 1990's and the early 2000's. These incidents have increased tension between India and Pakistan. In

FACTS

Official name:	Republic of India
Capital:	New Delhi
Terrain:	Upland plain in south, flat to rolling plain along the Ganges, deserts in west, Himalaya in north
Area:	1,269,219 mi² (3,287,263 km²)
Climate:	Varies from tropical monsoon in south to temperate in north
Main rivers:	Ganges, Krishna, Narmada, Yamuna, Godavari, Brahmaputra
Highest elevation:	Kanchenjunga, 28,208 ft (8,598 m)
Lowest elevation:	Indian Ocean, sea level
Form of government:	Federal republic
Head of state:	President
Head of government:	Prime minister
Administrative areas:	28 states, 7 union territories
Legislature:	Sansad (Parliament) consisting of the Rajya Sabha (Council of States) with a maximum of 250 members serving six-year terms and the Lok Sabha (People's Assembly) with 545 members serving five-year terms
Court system:	Supreme Court
Armed forces:	1,281,200 troops
National holiday:	Republic Day - January 26 (1950)
Estimated 2010 population:	1,202,135,000
Population density:	947 persons per mi² (366 per km²)
Population distribution:	71% rural, 29% urban
Life expectancy in years:	Male, 66; female, 69
Doctors per 1,000 people:	0.6
Birth rate per 1,000:	24
Death rate per 1,000:	8
Infant mortality:	54 deaths per 1,000 live births
Age structure:	0-14: 32%; 15-64: 63%; 65 and over: 5%
Internet users per 100 people:	7
Internet code:	.in
Languages spoken:	Hindi and 17 other official languages (Bengali, Telugu, Marathi, Tamil, Urdu, Gujarati, Malayalam, Kannada, Oriya, Punjabi, Assamese, Kashmiri, Sindhi, Sanskrit, Konkani, Manipuri, Nepali); English (associated language)
Religions:	Hindu 80.5%, Muslim 13.4%, Christian 2.3%, Sikh 1.9%, other 1.9%
Currency:	Indian rupee
Gross domestic product (GDP) in 2008:	$1.221 trillion U.S.
Real annual growth rate (2008):	6.6%
GDP per capita (2008):	$1,067 U.S.
Goods exported:	Chemicals, clothing, cotton, diamonds and jewelry, iron ore, machinery, petroleum products, transportation equipment
Goods imported:	Aircraft, chemicals, copper ore, crude oil, diamonds, gold, iron and steel, machinery
Trading partners:	China, Germany, Singapore, United Arab Emirates, United States

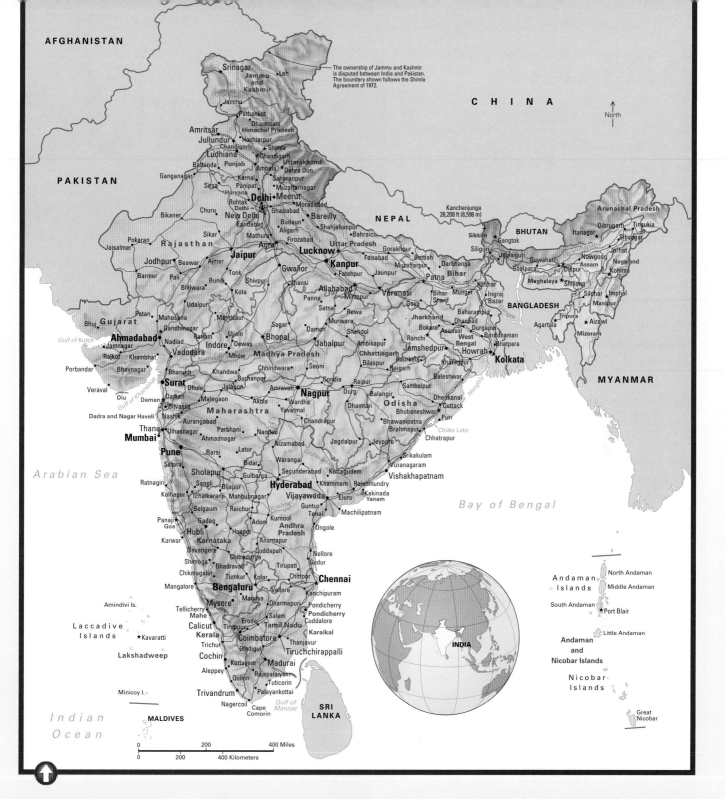

India extends into the Indian Ocean from its northern border in the Himalaya. The country's varied landscape also includes a desert, tropical rain forests, mighty rivers, and fertile plains.

2006, several bombs exploded on commuter trains in Mumbai, killing more than 180 people. In 2008, terrorists launched a series of attacks on Mumbai's main rail station, two hotels, a Jewish center, and a hospital, killing more than 170 people. Indian authorities accused Pakistani militants in both incidents.

India has also faced serious natural disasters. In 1993, an earthquake struck a rural area southeast of Mumbai, killing about 10,000 people. An earthquake in 2001 in the state of Gujarat killed tens of thousands of people. In 2004, a tsunami killed more than 16,000 people along India's eastern and southern coasts and in the Andaman and Nicobar Islands in the Bay of Bengal.

HISTORY

About 1500 B.C., a tribe of people called *Aryans* migrated to India from central Asia. They were the ancestors of the present-day northern Indians. In India, the Aryans found the Dravidians living in villages and growing crops. The Aryans gradually conquered the Dravidians and drove some of them south. The Dravidians were ancestors of the people who live in southern India today.

The influence of the Aryans gradually spread through all of India. The Aryans developed the Sanskrit language and established a way of writing it. Also during the time of Aryan rule, Siddhartha Gautama (563?-483? B.C.) founded the religion of Buddhism.

Empires and conquerors

Around 324 B.C., Chandragupta Maurya founded the Mauryan Empire, the first to unite almost all of India under one government. The most famous Mauryan ruler, Emperor Ashoka, helped spread Buddhism throughout the country.

About 185 B.C., the Mauryan Empire fell. Over the next 500 years, groups of central Asian people, including the Scythians and the Kushans, moved into northern India. About A.D. 320, native Indian emperors of the Gupta dynasty extended their empire across all of northern India and into what is now Afghanistan.

The golden age of India

The Gupta Empire, which lasted from about 320 to about 500, was the golden age of India. During this period, the arts flourished. Sanskrit, the classical language of Hindu religion and culture, flowered in poetry and literature. India's most famous dramatist and poet, Kalidasa, wrote works of great charm and beauty. The finest frescoes at Ajanta were also painted at this time, and many Hindu temples were built.

After the fall of the Gupta Empire, India was invaded by the Huns and other central Asian conquerors. In 1526, the Muslim ruler Babur invaded India and established the Mughal Empire. Babur's grandson, Akbar, ruled India wisely until his death in 1605, and he won the loyalty of the Hindus. However, Aurangzeb, who

Akbar
(1542-1605)

Mohandas K. Gandhi
(1869-1948)

Indira Gandhi
(1917-1984)

TIMELINE

c. 2500 B.C	Indus Valley civilization begins to flourish.
c. 1500 B.C	Aryans invade northern India.
400's B.C.	Siddhartha Gautama founds Buddhism.
324-185 B.C.	Maurya Empire unites India, peaks under reign of Emperor Ashoka (c. 272-232 B.C.).
c. A.D. 320	Gupta Empire brings a golden age c. 500 to India.
1206	Muslim sultanate established at Delhi.
1398	Timur raids India.
1498	Vasco da Gama of Portugal lands in India.
1526	Babur founds Mughal Empire.
1556-1605	Height of Mughal Empire under Akbar.
1600	British establish East India Company.
1757	Clive defeats Indian army at Plassey.
1774	Warren Hastings becomes first governor general of India.
c. 1800-1850	East India Company extends hold over Indian territories.
1857-1859	Indian soldiers lead the Indian Rebellion against British officials.
1858	British government establishes direct rule of India.
1885	Indian National Congress is formed.
1920	Mohandas K. Gandhi begins program of non-violent disobedience.
1947	India becomes independent on August 15, the day after Pakistan was created. Jawaharlal Nehru becomes India's first prime minister.
1947-194	India and Pakistan fight over control of Kashmir.
1948	Gandhi is assassinated.
1950	India's constitution takes effect.
1962	Chinese forces invade India over a border dispute, but pull back.
1965	India and Pakistan fight second war over control of Kashmir.
1966	Indira Gandhi, daughter of Nehru, becomes prime minister.
1971	East Pakistan becomes an independent nation called Bangladesh. India supports Bangladesh in civil war with West Pakistan.
1984	Indira Gandhi is assassinated. Her son Rajiv becomes prime minister.
1991	Rajiv Gandhi is assassinated.
1999	Heavy fighting takes place between Indian troops and Muslim guerrillas in Kashmir.
2004	Tsunami kills more than 16,000 in India.

became emperor in 1658, ruled harshly and alienated the Hindus. The Marathas of western India and other leaders in the south rebelled.

British India

With the death of Aurangzeb in 1707, there was no longer a central power in India. The East India Company of England, which had already set up trading posts in Bombay, Calcutta, and Madras (now called Mumbai, Kolkata, and Chennai), took advantage of this opportunity. In the mid-1700's, the company took control of much of India.

In 1857, Indian soldiers, called *sepoys,* rebelled against British officers who ordered them to bite open cartridges greased with fat from cows or hogs. The Hindu soldiers refused to obey the order because their religion did not allow them to eat beef. The Muslim soldiers refused because they were forbidden by their religion to eat pork. Soon, the Indian Rebellion spread through northern and central India. However, the rebels were poorly organized and had few weapons. By 1859, they had been defeated.

In 1858, the British government took direct control of all the land that had been governed by the East India Company. This area became known as *British India.* In most other parts of India, the British ruled indirectly through the local rulers. These areas were called the *princely,* or *native, states.*

The British built railroad, telegraph, and telephone systems. They also enlarged the Indian irrigation system. However, the Indian people criticized British policies and accused the British of holding them back in job opportunities and other areas. In 1885, with British approval, the Indian people formed the Indian National Congress to discuss their problems.

By 1920, Mohandas K. Gandhi had become a leader in the Indian National Congress and in the Indian independence movement. Gandhi defied British rule with a policy of *nonviolent disobedience,* which included nonpayment of taxes, sit-ins, and refusal to attend British schools. In 1947, the Indian people won their independence from the United Kingdom.

The East India Company established trading centers at Bombay, Calcutta, and Madras (now called Mumbai, Kolkata, and Chennai) in the early 1600's. By the early 1800's, the company controlled much of India. The government of the United Kingdom took control of the company's holdings in 1858.

Jawaharlal Nehru served as the first prime minister of independent India from 1947 until his death in 1964. Nehru worked to establish a democracy and to improve the living standards of the Indian people. His daughter, Indira Gandhi, and his grandson, Rajiv Gandhi, have also served as prime ministers.

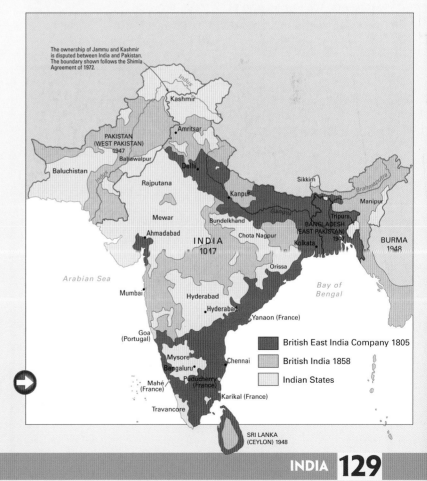

The ownership of Jammu and Kashmir is disputed between India and Pakistan. The boundary shown follows the Shimla Agreement of 1972.

Kashmir

PAKISTAN (WEST PAKISTAN) 1947

Amritsar

Bahawalpur

Baluchistan

Delhi

Rajputana

Sikkim

Kanpur

Mewar

Bundelkhand

Manipur

Ahmadabad

INDIA 1017

Chota Nagpur

Tripura

BANGLADESH (EAST PAKISTAN) 1947

Kolkata

BURMA 1948

Orissa

Arabian Sea

Mumbai

Hyderabad

Bay of Bengal

Hyderabad

Yanaon (France)

Goa (Portugal)

Mysore

Chennai

Bengaluru

Mahé (France)

Puducherry (France)

Karikal (France)

Travancore

SRI LANKA (CEYLON) 1948

British East India Company 1805

British India 1858

Indian States

ENVIRONMENT

India's land mass is shaped like a triangle, with its northern base in the Himalaya. Farther south, India extends into the Indian Ocean. The Arabian Sea borders its western edge, and the Bay of Bengal lies to the east.

India can be divided into three major land regions: (1) the Himalaya, (2) the Northern Plains, and (3) the Deccan, or Southern Plateau.

The Himalaya

The Himalaya mountain system extends along India's northern and northeastern border in a 1,500-mile (2,410-kilometer) curve.

The Himalaya is the tallest mountain system in the world. Kanchenjunga, India's tallest mountain, rises 28,208 feet (8,598 meters) high in the Himalaya on the Nepal border.

Because of the difference in altitude and exposure in many parts of the range, climate and plant life in the Himalaya are quite varied. Tropical plants, such as fig and palm trees, grow on the steep southern slopes up to 3,000 feet (910

Village women of the Northern Plains collect the droppings of domestic cattle, which they dry in the sun and use for fuel. Most of India's people live in the fertile Northern Plains.

Terraced fields extend down a slope in the Nilgiri Hills of southern India. The Nilgiris are the meeting point of the Eastern and Western Ghats on the Deccan, a vast upland area that covers half of India.

meters). At 12,000 feet (3,660 meters), cedar and other conifer trees flourish.

Many kinds of animals live in the Himalaya—from tigers and leopards to yaks, rhinoceroses, elephants, and some kinds of monkeys.

The Northern Plains

The Northern Plains, which lie between the Himalaya and the southern peninsula, include the valleys of the Brahmaputra, Ganges, and Indus rivers and their branches. The Northern Plains make up the world's largest *alluvial plain* (land formed of soil left by rivers). The soil in this region makes for excellent farmland, among the most fertile in the world. Irrigation is easy because the land is flat.

A highway winds through a tiny settlement in Himachal Pradesh toward a high pass in the Himalaya. These mountains separate India from China and the rest of Asia. The name Himalaya means House of Snow, or the Snowy Range, in Sanskrit.

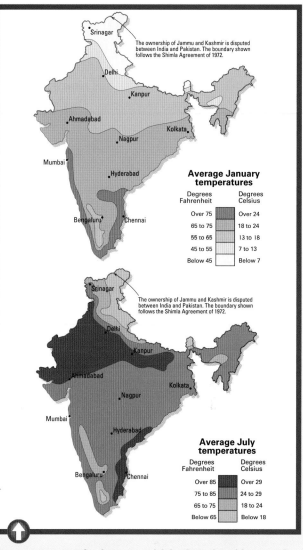

Temperature maps for January and July show the wide range in India's weather. The cool winds of the winter monsoon keep temperatures low in the northern areas, while in southern India, temperatures stay fairly high. By July, temperatures have already peaked, as the summer monsoon has brought rainfall. The hottest weather occurs between March and June. The Himalaya stays cool throughout the year, while the southern coasts have high temperatures all year long.

The Deccan

The Deccan, a huge plateau that forms most of the southern peninsula, is completely surrounded by mountains. To the north, the Vindhya and other mountain ranges separate the Deccan from the Northern Plains.

To the east, the Eastern Ghats rise along the edge of the Deccan and gradually slope to a wide coastal plain along the Bay of Bengal. The Western Ghats, on the western edge of the Deccan, fall sharply to a narrow coastal plain along the Arabian Sea. The Eastern and Western Ghats meet in the south in the Nilgiri Hills.

The Deccan includes farming and grazing land, as well as most of India's valuable mineral deposits. The forests along its eastern and western edges are the natural habitats of many large animals.

Climate

India's climate is quite varied. During the cool season (October through February), the weather is mild, except in the northern mountains where snow falls. The Northern Plains get the most intense heat during the hot season, which lasts from March to the end of June. The rainy season lasts from the middle of June through September, as the *monsoons* (seasonal winds) blow across the Indian Ocean from the southwest and southeast.

India's climate can be violent, with heavy rainfall, flash floods, extreme temperatures, and fierce winds that often occur without warning.

THE MONSOONS

The *monsoons* are seasonal winds that blow across the northern part of the Indian Ocean. They also blow across most of the surrounding land areas.

Monsoons are caused by the difference in the heating and cooling of air over land and sea. India has two monsoon seasons: the moist summer monsoon blows from the southeast and southwest from the middle of June through September, and the dry, cool winter monsoon blows from the northeast from November to March.

The summer monsoon is extremely important to agriculture in India because it brings much-needed rainfall. The winter monsoon brings very little rain because the cool air blowing from the Himalaya does not carry as much moisture as warm air.

Monsoons have an important effect on the nation's agriculture, so Indian meteorologists spend a great deal of time trying to understand and predict the behavior of these winds. Monsoons are known to be influenced by sea temperatures as far away as the South American coast. Other factors, such as a westerly *jet stream* (a fast-moving, high-altitude airstream)—or even the depth of snow on the Himalaya—may also affect the monsoons.

The summer monsoon

During the summer, the sun heats land surfaces far more than it does sea surfaces. As this strongly heated air over the land rises, it is replaced by a southwesterly wind carrying warm, moist air from the Indian Ocean. The winds from the ocean begin to rise as they move over land. The rising air cools, and the moisture in it condenses, forming clouds and rain.

Rain falls most heavily in northeastern India. Some hills and mountain slopes in this region receive an average of about 450 inches (1,140 centimeters) of rain per year. The world's heaviest recorded rainfall for a one-year period fell at Cherrapunji. This village had almost 1,042 inches (2,647 centimeters) of rain from August 1860 to July 1861.

People walk through the flooded streets of Mumbai during the summer monsoon. Indian people accept the threat of flooding and crop destruction in return for the life-giving rains of the summer monsoon.

Monsoons affect the pattern of rain and snowfall over much of Asia. Summer monsoons blow from the southeast and southwest from the middle of June through September, bringing much-needed rainfall to crops. Winter monsoons are dry winds that blow from the northeast from November to March.

A commuter train moves over flooded tracks in Mumbai following heavy rains. The monsoon usually covers the entire country by mid-July, providing the main source of water for agriculture.

Effect on agriculture

The arrival of the summer monsoon has an extremely important effect on Indian agriculture. Farmers depend on the rainfall brought by the summer monsoon to water their crops—if the monsoon brings enough rain, the crops will grow.

However, the monsoon does not always behave as expected. Sometimes the monsoon is late. Sometimes the monsoon rains start and then stop prematurely. And sometimes they turn back over the ocean. Crops then fail because they do not get enough rainfall.

A poor monsoon season creates other serious problems. Water reserves that are usually saved for later in the season are used up. Hydroelectricity plants are sometimes closed down, and farmers are left without electricity to run their well pumps. And sometimes the monsoon rains are so heavy that they cause serious flooding and crop destruction. Flooding in some areas and drought in other areas can occur in the same monsoon season.

In 1987, large parts of India suffered the worst drought in 40 years, while other areas saw the worst flooding in decades. The northwest state of Rajasthan suffered through a drought. Meanwhile in the northeast, Assam recorded the highest July rainfall in 50 years, and thousands of villages in Bihar suffered repeated flooding.

During much of India's history, failure of the monsoon could mean starvation and death for thousands of people in the affected areas. Although the threat of droughts and floods remains serious, modern transportation and other improvements have helped the country limit the level of suffering. India's extensive railroad system, for instance, has helped the government to transport food and assistance to villages hit by bad harvests.

Less than 10 inches (250mm)
10 to 40 inches (250 to 1.000mm)
40 to 80 inches (1.000 to 2.000mm)
More than 80 inches (2.000mm)

Kolkata

HIMALAYA MOUNTAINS
npur

Kolkata

Bay of Bengal

Chennai

AGRICULTURE

Agriculture is the largest and most important part of India's economy. Farms cover more than half of the country's area, and agriculture is the main source of income for the majority of India's people.

India's farmers grow a variety of crops, and India has the world's largest cattle population. However, because cows are sacred to Hindus, they are not butchered for meat. Farmers keep cows for milk production and for plowing. After the cows die, their hides are used to make leather.

Crop production

About three-fourths of India's farmland is used to grow grains and *pulses,* the country's main foods. Pulses are the seeds of various pod vegetables, such as beans, chickpeas, and pigeon peas. The major grain crops are corn, millet, rice, sorghum, and wheat.

India ranks as the world's leading producer of bananas, a fiber called *jute,* mangoes, millet, pulses, sesame seeds, and tea. It is a major grower of cabbages, cauliflower, coconuts, coffee, cotton, onions, oranges, peanuts, potatoes, rapeseeds, rubber, sugar cane, and tobacco. In addition, farmers grow such spices as cardamom, ginger, pepper, and turmeric.

Indian farms

Most farms in India are quite small. More than half are less than 2-1/2 acres (1 hectare) in area, and only a few cover more than 25 acres (10 hectares).

About two-thirds of India's farmers own their own land. However, many of these farms become smaller with each succeeding generation because of Indian inheritance customs. When a man dies, his land is divided up equally among his sons. In time, the property often becomes too small to cultivate profitably.

Most Indian farmers are *subsistence farmers,* who grow crops mainly to feed their families, not for commercial purposes. Large farms—such as those in the Punjab, which is called India's "breadbasket"— grow food for sale.

At harvesttime, women gather in the rice fields to cut the stalks with sickles. Rice is the staple crop in the eastern and southern regions.

India's farmers have two growing seasons: the main summer cultivation period, called the *kharif;* and the secondary winter season, the *rabi*. The kharif season produces the main harvest, provided the summer monsoon brings the proper rainfall.

Recent developments

In the past, India imported much of its food. Today, however, the country is essentially self-sufficient in food production—that is, it produces enough food to meet its needs. The increase in agricultural production came about partly because of the Green Revolution, the introduction of high-yielding seeds in the 1960's. Improved farming techniques, greater mechanization, and irrigation have also increased agricultural production. In addition, farmers are paid high prices for their crops to encourage them to grow more, and many rural development programs make credit and machinery easily available.

With the help of a pair of oxen, a farmer plows a wheat field near one of the famous Hindu temples in Khajuraho. Wheat is one of northern India's main foods. The people enjoy chapatties, thin, flat, baked breads made of wheat flour.

Tea plants grow throughout India, but the best-producing areas are in southern India and Assam. The well-known Darjeeling tea grows on hillsides near the city of the same name.

The richest agricultural regions of India are the Northern Plains, the Punjab, Gujarat, the Deccan, and the coastal regions. India now produces enough food to meet most of its needs. But floods and droughts still result in food shortages in some areas.

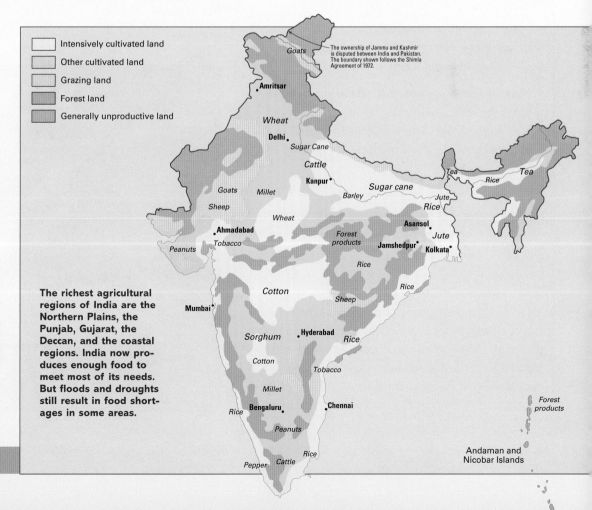

ECONOMY

India has a large economy, which is measured by its *gross national product* (GNP), the value of all goods and services produced in a year. But when the country's huge population is divided into its GNP, the GNP *per capita* (per person) is very low. For this reason, India is considered a developing country.

Since India became independent, the government has worked very hard to stimulate industrial growth. Although India is now one of the largest industrial nations in the world, the country still faces many challenges in its economic development. Most of its rich natural resources—farmland and mineral deposits—have not been sufficiently developed. And because the huge population keeps increasing so rapidly, it is difficult for economic gains to make a real difference.

Manufacturing

The clothing and textile industry is one of the largest employers in the country. Cotton mills are mainly located in western India, especially in the cities of Mumbai and Ahmadabad, and in southern India. Punjab has woolen mills, and Kolkata has jute factories.

India is one of the world's top producers of iron and steel. Huge iron and steel mills operate at Bhilai, Bokaro, Durgapur, and Raurkela.

Indian factories use iron and steel to make aircraft, automobiles, bicycles, electrical appliances, military equipment, and many kinds of industrial machinery. The assembly of electronic products is a growing industry in India today.

To continue its industrial growth, India must develop its natural resources further. Industry needs energy to develop, but India is at a disadvantage because it must import large quantities of oil.

Rich natural resources

India has valuable deposits of a variety of natural resources. The country is one of the world's leading producers of iron ore. Iron ore deposits lie mainly in Bihar and Odisha. India also has large deposits of coal and petroleum. Coal accounts for about 40 percent of the yearly value of all minerals mined in India, and petro-

Stockbrokers follow stock index numbers on their computer screens at a brokerage firm in Mumbai. The city is India's business, financial and trading capital.

India's motion-picture industry is the world's largest. This epic was filmed at the historic City Palace in Udaipur in the state of Rajesthan. All Indian popular films feature song-and-dance sequences.

leum accounts for about 30 percent. Huge coal deposits lie in Jharkland, Madhya Pradesh, Odisha, and the western end of West Bengal. There are some inland deposits of petroleum, mainly in Assam and Gujarat, but most drilling is off the shore of Mumbai.

India's mines produce a number minerals of useful for industrial development, including bauxite, chromite, gypsum, limestone, magnesite, manganese, mica, and titanium. The country also has deposits of precious metals and stones, including diamonds, emeralds, gold, and silver. Cut diamonds are one of India's biggest exports.

Cottage industries

Millions of Indian people throughout the country work at home, employed in what are called *cottage industries*. These workers make many beautiful items.

Home weavers create fine fabrics of cotton, rayon, and silk by hand, as well as beautifully designed carpets and rugs. They spin fine laces of gold and silver threads. Home craft workers also make brassware, jewelry, leather goods, and woodcarvings.

India's busy textile industry produces clothing made of home-grown cotton and jute, as well as wool, rayon, and colorful silks. The industry is one of the biggest sources of jobs in the country.

Nuclear power plants provide some of India's electricity. In 1963, the United States supplied the first such plant, near Mumbai.

Looking to the future

Service industries are economic activities that provide services, rather than produce goods. A growing urban population and increasing commercial and communication links between India and the rest of the world have led to a dramatic expansion of the country's service industries. Community, government, and personal services are the largest service areas. Other important service industries include finance, insurance, and real estate; and trade, restaurants, and hotels.

India's major stock exchange is in Mumbai, which is the nation's business, finance, and trading capital. Bengaluru is the center of the country's computer industry. Many multinational companies have customer-service call centers in India.

PEOPLE

For centuries, many different groups of people migrated to India from other parts of Asia. Today, the descendants of these ancient peoples give the population of India its great variety of ethnic and language groups.

The two largest ethnic groups are the Dravidians and the Indo-Aryans. The ancestors of the Dravidians were among the earliest known inhabitants of India. Today, most Dravidians live in the southern part of the country. Most Indo-Aryans live in the north. They are descended from the Aryans, an ancient people from central Asia, who invaded the area about 1500 B.C. The Aryans gradually conquered the Dravidians and drove some of them south.

Other central Asian groups settled in northern India from the A.D. 400's to 1400's. Their descendants are concentrated in the several northern states. Some groups who live in the far north and northeast are closely related to peoples of East and Southeast Asia.

A number of smaller groups of peoples live in remote forests and hills throughout India. Often referred to as tribes or tribal groups, these peoples include the Bhils, Gonds, Khasis, Mizos, Mundas, Oraons, and Santals.

Many languages

More than 1,000 languages and dialects are spoken in India. The major languages fall into two groups—the Indo-European and the Dravidian.

About three-fourths of the people speak the Indo-European languages, which include Hindi—India's most widely spoken language. Also included in this group are Urdu, which is closely related to Hindi; Assamese; Bengali; Gujarati; Kashmiri; Marathi; Oriya; Punjabi; and Sindhi. These languages come from Sanskrit, the ancient Indian language.

The Dravidian languages, spoken mainly in the southern part of the country, include Kan-

Men and women in Mumbai shop at a supermarket. The store is an example of how the West has made an impact on traditional Indian lifestyles.

Many Indian men wear turbans of various shapes and colors, especially during celebrations. These men have been taking part in the Hindu festival Holi by throwing colored powder and water on one another.

nada, Malayalam, Tamil, and Telugu. About a fifth of the population speaks these languages.

Two other language groups found in India are the Sino-Tibetan languages spoken in the northern Himalayan region and near the Myanmar border, and the Mon-Khmer languages spoken by some ethnic groups in northeastern and central India.

The principal official language of India is Hindi. Sanskrit and 16 regional languages are also official languages, while English is considered an "associate" national language. At least one major regional language is spoken in each state. Elementary and high school students study in their regional language and learn Hindi as a second language. In most colleges and universities, classes are taught in regional languages, but English is widely used.

Because so many languages are spoken in India, communication and understanding between ethnic groups has often been difficult. However, a more serious problem facing the Indian people today is overpopulation.

The population problem

India is the second most-populated country in the world. Overpopulation has caused serious overcrowding throughout India, particularly in the cities. In Kolkata, for example, the population density averages about 115,000 people per square mile (44,000 per square kilometer).

Millions of urban people live in slum dwellings. Often, an entire family inhabits a one-room shelter made of scraps of wood or metal. Many others are so poor that they have no home at all and must sleep in the streets.

India's schools are also overcrowded. Many children must drop out of school to help support their families. About two-thirds of India's adults can read and write. The lack of education limits job opportunities for many people, and, as a result, they continue to live in poverty.

To help control population growth, the government has introduced programs that encourage people to have smaller families.

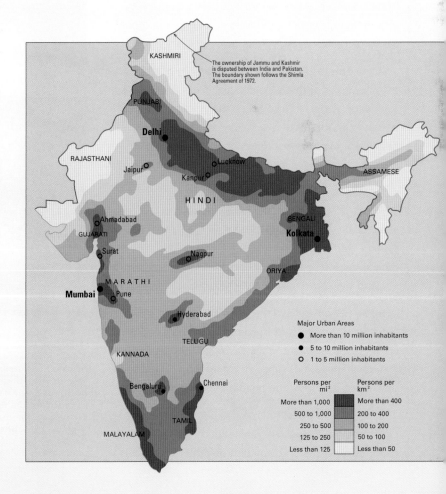

CULTURE

More than 80 percent of India's people practice the Hindu religion, one of the world's oldest living religions. Hindu rules and customs have an important influence on Indian life.

Unlike many other religions, Hinduism was not founded on the teachings of one person. Many different cultural, racial, and religious groups had a role in shaping Hindu philosophy.

The caste system

Indians, especially Hindus, have traditionally been organized into social groups called *castes*. Castes form the basis of a strict system of social classes.

Ancient Hindu texts described four main social groupings, called *varnas*. In order of rank, these varnas were (1) *Brahmans,* the priests and scholars; (2) *Kshatriyas,* the rulers and warriors; (3) *Vaishyas,* the merchants and professionals; and (4) *Shudras,* the laborers and servants. Over time, each varna came to include many smaller subdivisions, resulting in thousands of castes, each with its own rules of behavior.

According to Hindu belief, membership in a caste is established at birth and is difficult, if not impossible, to change. A person's social status in the community depends on the caste to which he or she belongs. Each caste also has a traditional occupation. Friendships and marriages rarely occur between members of widely different castes.

About 15 percent of the Indian population is considered outside the caste system. These people are called *untouchables,* or *scheduled castes.* They are ranked below the lowest Shudra caste in Indian society, and they have traditionally held the lowest jobs.

Some people believe that the caste system slows India's progress toward becoming a modern nation. In recent years, however, many caste barriers have broken down. People of different castes work in the same offices and factories and mingle in public places.

India's 1950 Constitution gave untouchables equal rights as full citizens. Discrimination in jobs and education against the untouchables is forbidden by law.

The Buddhist wallpainting from which this detail comes was painted in the cave temples of Ajanta, Maharashtra, during the 100's B.C. Buddhism was once the chief religion of India.

Hindu reverence for the cow dates from about 3,500 years ago. The sacredness of the cow is revealed in Hindu scriptures, particularly in the stories of Krishna among the gopis (milkmaids) of Brindaban. Krishna is a Hindu god. His conversations with the Pandava warrior Arjuna discuss the meaning and nature of existence in the philosophical work Bhagavad-Gita. Reverence for cows is also part of the Hindu philosophy of ahimsa, noninjury to living creatures.

Ganesha, the elephant-headed Hindu god, is featured in this Tamil Nadu temple shrine. Ganesha is universally honored by Hindus as the "remover of obstacles." Hindus worship many divinities (gods and goddesses).

Hindus believe that animals as well as human beings have souls. They have reverence for cows, monkeys, and other animals. Although not all Hindus are vegetarians, most will not eat beef.

Minority faiths

Most Indians are Hindus. However, Muslims, Christians, and other religious groups also live in India.

About 13 percent of the Indian people are Muslims, making Islam the second largest religion in India. Most Muslims live in the northern part of the country. Many of India's Christians live in the southern states of Kerala and Tamil Nadu and in the areas along India's northeast border.

Buddhism and Jainism were founded in India in the 500's and 400's B.C. A small percentage of the population belongs to each of these religions. The Sikh religion was founded in northern India by a *guru* (teacher) named Nanak around A.D. 1500. The majority of Sikhs live in the northwestern Indian state of Punjab. Members of a number of other religions also live in India.

The government has also set aside for them and other disadvantaged groups a significant percentage of government jobs, scholarships, and legislative seats. Nevertheless, the untouchables remain an oppressed group, especially in villages.

Hindu worship

Hindu belief and conduct is based on the teachings of Sanskrit literature. *The Vedas* are the oldest Hindu scriptures. These philosophical works existed for centuries and were passed down orally from generation to generation before they were written down.

Many ancient Hindu rituals are still widely observed. Millions of Hindus visit temples along the Ganges River, the most sacred river in India. Hindu temples hold annual festivals to honor events in the lives of their gods, and most Hindus have a shrine in their home devoted to a particular god.

The Hindu Dussehra festival, celebrated here in Delhi, commemorates events in the lives of the divinities. These festivals attract huge crowds of Hindus, who come to worship, to pray, and to enjoy the colorful display.

THE RIVER GANGES

The Ganges is the most important river in India and one of the largest in the world. It begins as a pool in a Himalayan ice cave 10,300 feet (3,139 meters) above sea level.

The river flows southeast through the northern part of India and into Bangladesh. There, some of the branches of the Ganges join the Brahmaputra River to form a large delta, and together they flow into the Bay of Bengal.

The Ganges is one of India's greatest natural resources. As the river flows through the country, it leaves soil deposits that make the land rich and fertile for farming. Through irrigation systems, its waters provide moisture for India's crops.

In the past, when the Ganges River carried many merchant ships, some of India's largest cities were built along its banks. In recent years, irrigation has drained much of its water, and steamboats can navigate only in the lower river. As a result, the river has become less important for trade.

To the people of India, however, the Ganges—or Ganga, as the Hindus call it—has far more importance for its spiritual value. Hindus consider it the most sacred river in India.

The goddess Ganga

Ancient Hindu scriptures tell how, for a very long time, the goddess Ganga flowed only in the heavens. She was an object of great beauty and sacredness, but she was of little use to the dry earth below.

The Ganges flows through a steep-sided valley in its final course through the Himalaya. Near Rishikesh, the river leaves the mountains and flows to the Northern Plains by way of the city of Howrah.

Hindu holy men meditate at the source of the River Ganges in the Himalaya. These men have renounced all worldly attachments and live in caves, forests, and temples throughout India and Nepal.

The Maharaja Sagar, an Indian king of great riches, had 1,000 sons who traveled far and wide in search of new lands to conquer. During their journey, they came upon a Hindu wise man named Kapil. Annoyed because the sons had disturbed his meditation, Kapil burned them to death with his eyes.

When the maharaja heard what had happened to his sons, he begged Kapil to return them to life. Kapil agreed to restore the sons' lives, but the goddess Ganga would have to come down from the heavens and touch their ashes.

Ganga was afraid that her mighty torrent of water would shatter the earth's foundation. She asked the god Shiva to stand above the earth on the rock and ice of the Himalaya.

Ganga then flowed down to earth through Shiva's matted hair, which absorbed the shock of her waves.

She trickled down from the mountains and across the plains, bringing water and life to the dry earth. Ganga then touched the ashes lying there, and the maharaja's sons came back to life. Thus, the earthly Ganga was born.

Ritual bathing

The holy cities on the banks of the Ganges attract thousands of Hindu pilgrims. They bathe in the river and take home some of its water. The river is said to bring purity, wealth, and fertility to those who bathe in it. The Hindus have built *ghats* (stairways) along the banks of the Ganges in order to reach the water more easily.

Some pilgrims bathe in the Ganges to cleanse and purify themselves. The sick bathe in the Ganges in the hope that the sacred waters will cure their illnesses. Others come to die in the river, because the Hindus believe that those who do will be taken to Paradise.

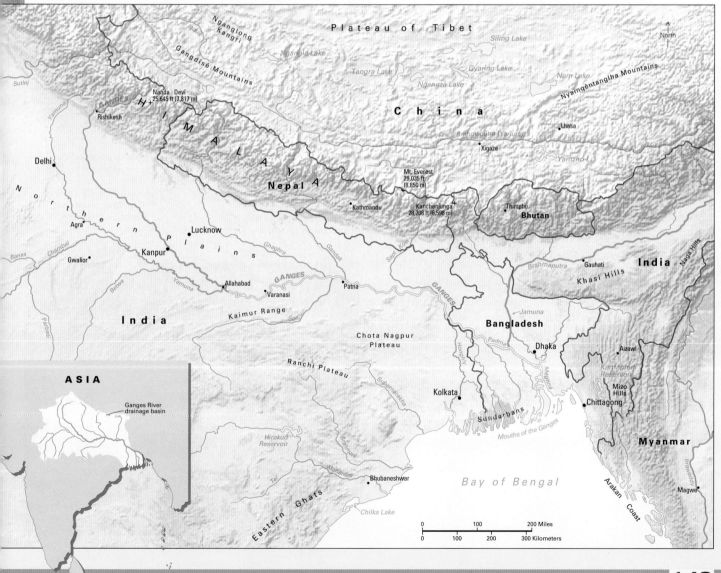

CITIES AND VILLAGES

India has been described as a land of villages. About 71 percent of its people live in villages. Most villages are small, and the people work mainly as farmers.

Since the mid-1900's, the population of India's cities has grown dramatically. Millions of villagers have left the rural areas to look for work in cities, where wages are higher. The largest cities are Ahmadabad, Bengaluru, Chennai, Delhi, Hyderabad, Kolkata, and Mumbai.

Village life

Indian villages, like the people who live in them, look quite different from one part of the country to another. The stone dwellings of the chilly mountain regions are quite unlike the bamboo and matting houses of the hot central and southern areas. Some villages are small settlements around isolated farms, while others are tightly knit communities built around larger farms.

Most villagers live in clusters of dwellings made of mud and straw. These houses usually have mud floors and only one or two rooms. Household articles may include brass pots for cooking, clay pots for carrying water and storing food, and little else. If a village or some of its houses are without electricity, people use kerosene lanterns.

Many village homes have no running water. The women get water from the village well, often the center of activity. Sometimes the water source is a pond or nearby river. The women pour the water into pots and carry the pots home on their heads.

Meals usually consist of rice and *dal,* a porridge made of *pulses,* which are seeds of such pod vegetables as beans, chickpeas, pigeon peas, and lentils.

Family ties are important to the villagers. Many people regard marriage as a relationship between two families, rather than a union between two individuals. An Indian household may include not just parents and children, but also the sons' wives and their children. Relatives and neighbors often join together to help those who have met with some misfortune.

Rural life has improved in many regions. Economic growth has brought such modern conveniences as electricity, improved drinking water and sanitation, and paved roads.

In Chandni Chowk, or Silver Square, the commercial center of Old Delhi, merchants sell silver jewelry, wholesale goods, sweetmeats, handicrafts, and clothing in market stalls.

An artificial reservoir, known as the village tank, is often the heart of small communities throughout India. Many were built centuries ago and may be a village's only source of water.

Village children attend school at Naricanda in the Himalayan foothills. Government education programs have improved the country's literacy rate, and now about two-thirds of India's adults can read and write.

In the slums of Mumbai, many people live in shacks made of wood or metal scraps. Others are crowded into high-rise tene-ment buildings. The slum areas have poor water supplies and sanitation.

City life

India's oldest cities were founded in ancient times. Today, they have densely populated city centers, often within the areas once surrounded by protec-tive city walls. Buildings occupy most of the space in the city centers, where twisting streets bustle with activity.

When the United Kingdom ruled India, the British people built and generally lived in sepa-rate areas called *cantonments,* which had wide streets and were less crowded. Today, Indian politicians, military officers, wealthy business peo-ple, and other leaders live in the cantonments.

Most Indian cities have a growing middle class, including government employees, office workers, and shopkeepers. Many urban dwellers work at manual labor or in factory jobs.

The huge growth of Indian cities since the mid-1900's has strained the resources of many cities, making it difficult to provide enough electricity, housing, and water for all the inhabitants. Millions of people live in slums, where water supplies and sanitation are poor.

THE HOLY CITY OF VARANASI

Along a sandy ridge on the west bank of the Ganges River lies the city of Varanasi, also known as Banaras, Benares, or Kashi. One of the largest cities in the northern Indian state of Uttar Pradesh, Varanasi is known for the fine silk fabrics made by its craft workers. For India's Hindus, however, Varanasi is a place of deep religious importance.

Holy bathing

The Hindus believe that the whole length of the Ganges is sacred, and that Varanasi is the holiest place on earth. Each year, about 2 million Hindu pilgrims from all over India come to the city to bathe in the waters. Many also come to Varanasi to die, because Hindus believe that those who die in the Ganges will be carried directly to Paradise.

Pilgrims enter the Ganges from one of the more than 80 *ghats* (steps leading down to the river) in Varanasi. The ghats extend about 5 miles (8 kilometers) along the riverbanks. Each day, before the light of dawn, devout Hindus begin to gather for their holy bath.

Three ghats are reserved for cremating the dead. More than 3,000 cremations take place every year in Varanasi at the ghats of Jalasayi, Harish Chandra, and Manikarnika.

The ancient ghat called Manikarnika takes its name from the *manikarnika* (earring jewel), which belonged to Sati, wife of the Hindu god Shiva. According to Hindu scriptures, Brahman priests of the highest caste found Sati's lost earring but did not return it. Shiva punished the Brahmans by lowering their caste status to those of *doams*. He put the doams in charge of the burning ghats. Today, the doams still tend the funeral pyres at the burning ghats.

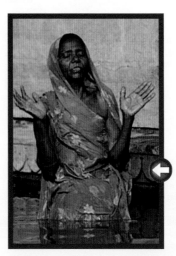

Hindus believe that bathing in the Ganges cleanses the soul. Pilgrims come to Varanasi from all over India to perform this ritual.

The holy bathing ghats of Varanasi date from the A.D. 300's. The ghatia priests, with their huge straw umbrellas, await the faithful.

Hindu pilgrims flock to the city of Varanasi to purify themselves in the waters of the Ganges. Devout Hindus expect to come to Varanasi at least once in their lives.

⬆

A funeral pyre goes up in flames at a burning ghat while another body is prepared for cremation. The task of cremation is given to the doams, a Hindu caste.

Religious and cultural center

Varanasi is sacred to the Buddhists as well as to the Hindus. It was in Sarnath, just a few miles from Varanasi, that the Buddha preached his first sermon to five holy men. During his sermon, the Buddha preached the message of how to overcome suffering. The delivery of this sermon is one of Buddhism's most sacred events, and Varanasi is one of Buddhism's most sacred places.

The city is also a center of culture and learning. Banaras Hindu University and other colleges are located there. Varanasi is also the site of a library containing thousands of ancient manuscripts and other documents.

Scholars and sages, from Shankaracharya (700?-750?) to Vivekananda (1862-1902), have found inspiration in Varanasi. The city is also the birthplace of Ravi Shankar, a world-famous composer and sitarist. A *sitar* is a stringed instrument used to play the classical music of India.

The arts and crafts of Varanasi have been prized since ancient times. In addition to their famous silk fabrics, the city's craft workers make shawls, *saris* (long dresses worn by Indian women), gold-embroidered cloth, hand-hammered brassware, and heavy gold and silver jewelry.

EXPLORING INDIA

India has fascinated Westerners for centuries. To early European explorers, traders, and adventurers, India was a land of mystery and excitement. Today, visitors speak of its timeless magic, where past meets present at every turn.

It is a land of great variety in people, cultures, and landscapes. In no other country would a visitor be greeted by a hotel clerk wearing a *dhoti* (a simple white garment wrapped between the legs) and displaying a caste mark on his forehead. India is also the only country in the world where a traffic jam could be caused by a sacred cow deciding to sit down in the middle of a city street.

Indian leaders understand the value of tourism to India's economy. In recent years, they have improved tourist facilities and opened up new areas of the country to visitors.

Because India is so vast, a tour of one to three weeks is not long enough to see even a large part of the country. As a result, most tours take in only a single region. Even so, visitors have a variety of places from which to choose. They may enjoy the glamour and bustle of city life in Mumbai or Kolkata or the sun-drenched beaches of Goa.

Northern India is most enjoyable in the cool, mild months of October through February. In southern India, where it is almost always hot, November through January is the coolest time.

Wherever tourists go in India, they are welcomed by the generous nature and friendliness of the Indian people. Visitors are advised to return the traditional Indian greeting of *namaste*—placing palms together in front of the chest and bowing. The Indian people prefer not to shake hands.

For visitors to northern India, the city of Delhi is a good place to start. Delhi is India's second largest city and features many historic landmarks.

The Red Fort, one of the city's most impressive monuments, is a red sandstone structure that covers several blocks. It was built between 1639 and 1648 by the Mughal Emperor Shah Jahan. The remains of the imperial palace and other Mughal structures lie within its walls.

While the Mughal Empire left its mark on Old Delhi, the British influence can be seen in New Delhi. Now the capital

The state of Tamil Nadu, in southern India, has many Hindu temples. Hindu sculpture dating from the 600's may be seen at the Seven Pagodas in Mahabalipuram, south of the seaport city of Chennai.

Few Indians own automobiles. They travel on trains and buses or ride motorcycles and bicycles. India's railway system, which is owned and operated by the government, is one of the world's largest.

The lake palace in Amber, just outside Jaipur in the northwestern state of Rajasthan, was built by Mughal rulers in the 1600's. Its peaceful setting on a steep hill skirted by a lake shows the beauty of rural India.

Carved from very hard sandstone, the detailed latticework of the Maharaja's Palace in Jodhpur shows the great skill and patience of India's traditional stonemasons. The palace, one of the largest residences in the world, was built between 1929 and 1943.

of India, New Delhi was built in the early 1900's, when India was part of the British Empire. It features wide, treelined avenues and many gardens.

South of Delhi lies the city of Agra, the site of India's most famous monument—the Taj Mahal. One of the most beautiful and costly tombs in the world, the Taj Mahal was built by Shah Jahan. Construction lasted from about 1630 to 1650.

The island city of Mumbai, formerly called Bombay, is situated on the west coast of India. In its harbor stands Elephanta Island, with its cave temples dating from the A.D. 600's to 900's. East of Mumbai, near Aurangabad, the Buddhist monasteries and shrines in the caves of Ajanta are decorated with *frescoes* (wallpaintings) from the Gupta period.

Southwest of Chennai in Tamil Nadu stands the sacred city of Kanchipuram. A Dravidian political capital as early as the A.D. 500's, Kanchipuram is the site of more than 100 Hindu temples.

PROJECT TIGER

At the beginning of the 1900's, about 40,000 tigers roamed the jungles and grasslands of India. In 1972, the Indian government's first formal tiger census found less than 2,000 tigers left in the entire country. Once common throughout southern Asia, the tiger had become an endangered species.

The low population of the tigers alarmed conservationists and wildlife experts. The year after the tiger census, Prime Minister Indira Gandhi launched "Project Tiger," thought by many to be the most dramatic rescue operation in conservation history.

Project Tiger began in 1973 with the formation of nine tiger preserves. The first reserve was organized at the Corbett National Park. Corbett, originally founded as India's first national park in 1936, is located in the northern state of Uttarakhand.

The tiger preserves provide enough space for a natural tiger habitat, while also protecting the human population who live in the area. By the early 2000's, 27 tiger reserves had been established.

Project Tiger has been very successful. According to official estimates, the tiger population in India has doubled since the program began. Yet the threat to tigers continues. After the program's early success, tiger numbers began to decline again. Poachers began snaring and poisoning tigers. It was reported that tigers were being killed so that their bones could be used in medicines. By the early 1990's, it had become increasingly difficult for tourists and trackers to spot tigers, and conservationists renewed their fund-raising and publicity efforts.

Threatened with extinction

Several factors have contributed to the drop in India's tiger population. Farmers have cut down forests to make room for grazing land and to obtain wood. As a result, tigers had to compete for living space with millions of Indian people.

Hunting has also contributed to making the tiger an endangered species. A government ban on shooting tigers went into effect in 1970.

In 1973, an international treaty called the Convention on International Trade in Endangered

A tiger cools off in the water. Tigers often soak themselves for an hour or so in marshes and rivers. Tigers are good swimmers and may swim across rivers in search of prey.

Scientists attach a monitoring device to a tranquilized female tiger. These devices are used to track the movements of animals in Project Tiger reserves, allowing scientists to learn more about their habits.

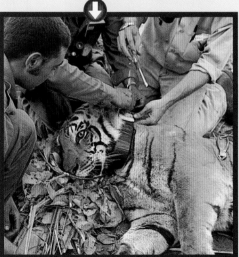

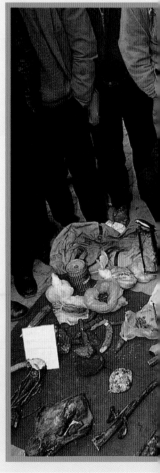

where nature is allowed to take its own course. Surrounding the core area is a larger *buffer zone*, which people are allowed to enter for a specific reason, such as to cut bamboo.

Saving many animals

The success of Project Tiger is important for many reasons. The tiger has been saved from extinction—at least for now. Other species also benefit by the preservation of the tiger's large natural habitat.

India has a wealth of magnificent wild creatures. Many of these, including the elephant, the Indian rhinoceros, the gaur, and the barasingha, are rare or endangered species. Some of these species now live in Project Tiger reserves with no interference from human activity.

Project Tiger has also taught people around the world that humankind and nature can—and must—live in harmony.

The illegal sale of tiger parts continues, despite increasing awareness of the tiger's plight.

Species of Wild Fauna and Flora (CITES) was drawn up. The treaty forbids commercial international trade in skins and animal parts of certain endangered species, including tigers. Nine nations originally became parties to the treaty in 1975. Since then, more than 170 nations have signed on to it. Nevertheless, illegal trade in many endangered animals still continues.

Creating the reserves

Because tigers are large predators, they need a great deal of land for their habitats. Therefore, the leaders of Project Tiger created large reserves that give tigers enough room to live, hunt for prey, and raise their cubs.

Each reserve consists of a *core area*, which people are forbidden to enter and

A monkey becomes a meal for a tiger in the Corbett National Park. Although they prefer large prey, tigers will also feed on such small prey as peafowl, monkeys, tortoises, frogs, and porcupines.

SIKKIM

Sikkim is the second smallest Indian state. It covers an area of 2,740 square miles (7,096 square kilometers) tucked away high in the Himalaya. Sikkim is north of the Indian state of West Bengali. Nepal lies to the west of Sikkim, China to the north and east, and Bhutan to the southeast.

Before Sikkim became a state of India in 1975, it was a monarchy ruled by a *chogyal* (king). Sikkim first became an independent monarchy about 1640, when Penchu Namgyal was crowned chogyal. At that time, Sikkim controlled lands that are now part of Bhutan, China, India, and Nepal.

In 1780, warriors from Nepal and Bhutan invaded Sikkim and seized much of the land. The United Kingdom returned some of this land to Sikkim when it conquered the Nepalese in 1814.

Sikkim came under the protection of the United Kingdom in 1861. Later, a British official took over much of the chogyal's power. By 1918, the chogyal had regained control over internal matters.

The United Kingdom gave control of Sikkim to India when India became an independent nation in 1947. The Indo-Sikkim Treaty of 1950 gave India much of the same powers that the United Kingdom had before 1947. Sikkim was independent in its internal affairs, but India controlled its foreign relations, defense, and communications systems.

In April 1975, a special referendum was held in Sikkim, and voters approved the legislature's proposal to become a state of India. In May, Sikkim officially became an Indian state.

Landscape and economy

Although Sikkim is relatively small in area, its landscape is extremely varied. Towering mountains extend in an arc along its western, north-

Lamaist monks often enter the order when they are very young. They are devoted to the teachings of Buddha, the founder of Buddism who lived in the late 500's to early 400's B.C.

Tiny Sikkim lies on India's northern border with China. It is nestled between the Himalayan kingdoms of Nepal in the west and Bhutan in the east.

ern, and eastern borders. Mount Kanchenjunga, the third highest mountain in the world, lies on the western border with Nepal.

The southern border opens to a broad valley watered by the Tista River, the state's major waterway. In the river valleys, thick tropical rain forests cover the land.

Farming is the basis of the Sikkim economy. Crops include rice, corn, and other cereals. Apples, cardamom, citrus fruits, pineapples, and potatoes are also grown.

Handicrafts are the chief Sikkim industry. Skilled craftworkers weave cloth, blankets, and rugs. Sikkim is also known for its copperware and woodcarvings.

Way of life

Most of the people who live in Sikkim are Nepalese, Lepchas, and Bhutias. The Nepalese, who make up about 70 percent of the population, live mainly on small farms in the southern regions. They work the land with simple hand tools. Most Nepalese speak Nepali. They are Hindus, but their religion is strongly influenced by Buddhism.

The Lepchas, the first settlers in Sikkim, now live in distant valleys, hunting and fishing for food. Some Lepchas also farm and raise livestock. Most speak Sikkimese.

The Bhutias are a seminomadic people. In summer, they live in tents and herd cattle and *yaks* (Asian oxen) in the high mountain meadows. In winter, they live in wooden houses in the highlands. Like the Lepchas, the Bhutias speak Sikkimese. Both groups practice Lamaism, also known as Tibetan Buddhism.

Grassy meadows surround Gangtok, Sikkim's capital and its only city. Gangtok is located in the southeast part of the state along the route used for centuries by major caravans between Tibet and India.

Religious processions are a common sight in Sikkim. The Lepchas and Bhutias practice Lamaism, a form of Buddhism. The Nepalese of Sikkim are Hindus.

GOA

In a small area of the west coast of India, the clear blue waves of the Arabian Sea roll slowly over the white sands of a sun-drenched beach. Warm sea breezes blow inland across the rooftops of an old fishing village. This is Goa, a tiny paradise of sandy beaches, sleepy villages, and warm, tropical weather.

Now an Indian state, Goa was formerly a Portuguese colony and later a territory of India. It covers an area of about 1,429 square miles (3,702 square kilometers). It lies about 250 miles (400 kilometers) south of Mumbai, between the Western Ghats and the Arabian Sea.

Early history

The ancient Hindu city of Goa was famous for its beauty, wealth, and culture. References to Goa appear in the early Hindu epic poems, the *Ramayana* and the *Mahabharata*.

Goa's political history can be traced back to the 200's B.C., when it was part of the Mauryan Empire. Next, a series of small kingdoms ruled the region, which became an important trading center.

The Portuguese were the first Europeans to appear on the shores of Goa. Alfonso de Albuquerque arrived in 1510. He conquered the area and soon controlled the coast. The region became a powerful and wealthy trading area, where silks, spices, porcelains, and pearls passed in and out of Goa's harbors. The city became known as *Golden Goa*.

Thirty-two years after Albuquerque arrived in Goa, Saint Francis Xavier came to Goa as a Christian missionary. Although Saint Francis Xavier left Goa to preach in distant lands, his body was returned after his death in 1552. He is buried in the *Bom Jesus Basilica* in the city of Old Goa. Today, most Goans who live in the coastal region are Roman Catholics, and most inland inhabitants are Hindus.

Goa reached the height of its power in the 1500's. During the 1600's to early 1700's, the Dutch and the Marathas of western India attacked, but they failed to capture Goa. Meanwhile, cholera outbreaks devastated the city of Old Goa, and the

Goa is a former Portuguese colony on the west coast of India. It is now an Indian state. About 1.35 million people live in Goa, which is known for its fine beaches and warm, tropical climate. Its capital city is Panaji.

The beach at Baga, Goa, is a beautiful backdrop for an outrigger fishing boat at sunset. Fresh fish are a favorite dish in many local hotels and restaurants.

Many young travelers come to Goa for its relaxed atmosphere. Because Goa's economy depends on tourism, many tourist facilities are available.

The highly decorated church of Saint Cajetan was modeled after St. Peter's Basilica in Rome. The church's crypt (underground chamber) holds the tombs of generations of Portuguese rulers. The church features a large arched window with tiny panes made of sea shells.

seat of government was moved to Panaji in 1759. The Portuguese controlled the region until the mid-1900's.

When India became independent in 1947, the government tried to persuade the Portuguese to withdraw from Goa. In 1961, the Indian army invaded Goa and ended 451 years of colonial rule. Goa was first declared a Union Territory along with the tiny colonies of Daman and Diu. In 1987, Goa became a separate state of India.

Goa today

Most Goans work in agriculture. Rice is the main crop, and Goans produce high yields on their irrigated fields. Farmers export bananas, cashew nuts, coconuts, mangoes, pineapples, pulses, spices, and sugar cane. Processing cashews is also an important industry.

Mormugao is one of India's main commercial seaports. Goa is also the base for more than 4,000 fishing boats. Fish is an important food and a major export. Goa also mines and exports bauxite, iron ore, and manganese. The state's beautiful tropical beaches and historic buildings make it popular with tourists.

Rural Goan families suffered great hardship under Portuguese rule. Conditions have improved since India took control of Goa in 1961. Today, government programs provide better housing and more effective health care.

INDIAN ISLANDS

In addition to its huge peninsula, the country of India also includes two island territories. These are (1) the Union Territory of Lakshadweep and (2) the Union Territory of the Andaman and Nicobar Islands.

The islands of Lakshadweep rise in the Arabian Sea about 125 miles (200 kilometers) off the southwest coast of India. The Andaman and Nicobar Islands are situated in the Bay of Bengal, 740 miles (1,190 kilometers) off the southeast coast of India.

Lakshadweep

Even though the name *Lakshadweep* means *thousand islands*, the group consists of 12 small coral atolls and some reefs, with a total land area of 12 square miles (32 square kilometers). About 60,600 people live in the territory on 10 inhabited islands.

Most of the islanders belong to various Arabian tribes. They speak the Malayalam language, a Dravidian language commonly used on India's southwest coast.

The island women make an elastic fiber called *coir* from coconut husks. Coir is used in the manufacture of matting. The men build boats and trade coir, processed fish, palm oil, and vegetables for rice. Rice is their staple food.

In the past, the islands of Lakshadweep were India's western outpost and served as a navigation point between Africa, Arabia, and India. Today, the islands are slowly opening up to tourism.

The Andaman and Nicobar Islands

Like Lakshadweep, the Andamans and the Nicobars have the status of Union Territory. They have been called *Marigold Sun* for their natural beauty. The 223 islands dot an area in the Bay of Bengal covering about 3,185 square miles (8,250 square kilometers).

Of these 223 islands, 204 belong to the Andamans, including the "big" islands of North, Mid-

Fishing boats and nets form a pattern on the skyline at Cochin on India's mainland. Cochin is the main port of call for traders from the Lakshadweep Territory. The islanders bring coconut fiber and mats to trade for rice.

dle, South, and Little Andaman. The largest of the Nicobar islands are Greater and Lesser Nicobar, Camorta, Katchall, and Car Nicobar. The capital, Port Blair, in the Andaman Islands, is the only town. It has about 106,000 people. Most of the Andaman and Nicobar Islands are uninhabited.

The landscape of the islands is marked by high mountain peaks and dense rain forests. Only the narrow coastal plains, valleys, and the mouths of rivers are suitable for settlement. Because of the great demand for land on the islands, forests are being cleared at a rapid rate.

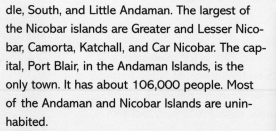

The Andaman and Nicobar Islands enjoy the protection of the Indian government as a Union Territory of India.

ASIA
China
Pakistan
Bhutan
Nepal
Bangladesh
India
Myanmar
Bay of Bengal
ANDAMAN IS.
Sri Lanka
NICOBAR IS.
Maldives
Indian Ocean

The Union Territory of Lakshadweep consists of the Laccadive Islands and Minicoy Island.

ASIA
China
Pakistan
Bhutan
Nepal
Bangladesh
India
Arabian Sea
LAKSHADWEEP
Sri Lanka
Maldives
Indian Ocean

This engraving from the 1800's shows a paddle steamer approaching the Andaman Islands while islanders attempt to defend their land from invasion. Port Blair on South Andaman was a British penal colony from 1858 to 1945.

The people of the Andaman and Nicobar Islands belong to a variety of ethnic groups. Some Indian islanders are the descendants of convicts brought there in the late 1800's and early 1900's, when Port Blair was a British penal colony.

Migrants and refugees from East Pakistan (now Bangladesh) also settled on the islands. The arrival of these outsiders greatly reduced the numbers of original inhabitants, who belonged to several ethnic groups. Some members of these groups have mixed with the later arrivals, while others survive in forest reservations on the main islands or in voluntary isolation on the islets and in the mountains.

The islanders grow rice, coconut, rubber, coffee, and tropical fruits and vegetables. Teak and structural timber are the islands' most important exports.

In December 2004, a tsunami killed about 1,900 people and caused enormous damage in the Andaman and Nicobar Islands.

Very few descendants of the original Andaman Island inhabitants remain today. Some of the survivors live by hunting and fishing. They also collect honey from the wild bees in the islands' dense tropical rain forests.

JAMMU AND KASHMIR

Jammu and Kashmir is a state in the far north of India. It is made up of three territories—Jammu, Kashmir, and Ladakh. Jammu has a largely Hindu population. Kashmir is predominantly Muslim. Ladakh's population is divided between Buddhists and Muslims.

Jammu and Kashmir is a land of towering mountain peaks, dense forests, deep blue lakes, and well-watered river valleys. Its summer capital and largest city, Srinagar, is known as the *Venice of the East* because it is crisscrossed by many canals. It was once the summer resort of India's Mughal rulers. The mountains of the Karakoram and Himalaya systems cut diagonally across the land. K2, the world's second tallest mountain peak, rises 28,250 feet (8,611 meters) in the Karakoram range.

The Indus River separates the Himalaya and the Karakoram mountains as it flows northwestward through Kashmir. Its main tributary, the Jhelum River, flows through the famous Vale of Kashmir.

Jammu and Kashmir's heartland

The Vale of Kashmir extends about 85 miles (140 kilometers) from northwest to southeast. It is about 20 miles (32 kilometers) wide. The surrounding mountains protect the vale, which enjoys moderate rainfall and long, warm summers.

The mild climate and well-watered soil of this river valley have encouraged the growth of agriculture. The vale is famous for its *saffron*. This yellow dye and food flavoring is obtained by drying the stigmas and part of the style of the purple autumn crocus.

Rice is the major crop in Jammu and Kashmir. It is grown on the plains where the paddies are watered by canals and on irrigated terraces on the lower mountain slopes. Corn and other grains are grown in the drier areas. Orchards produce apples, cherries, oranges, peaches, and pears for export. Roses and jasmine provide oil used in perfumes. A unique feature is the vegetable gardens floating on Kashmir's many lakes.

Many farmers also raise livestock. Flocks of Cashmere goats graze on the cold, high, dry meadowlands of Ladakh. Underneath their top coat of hair, Cashmere goats have soft fleece that is used to produce *cashmere,* a fine silky wool.

Jammu and Kashmir is known for wood products and silk, and especially for beautiful rugs and shawls made from cashmere and silk. The government of Jammu and Kashmir is making efforts to attract information technology and electronics industries to the state.

Farms are perched on the sides of hills in a remote river valley in Kashmir. Rice and corn are the main crops, but only 6 percent of the land is suitable for farming. Cedar and pine are important resources from Kashmir's dense forests.

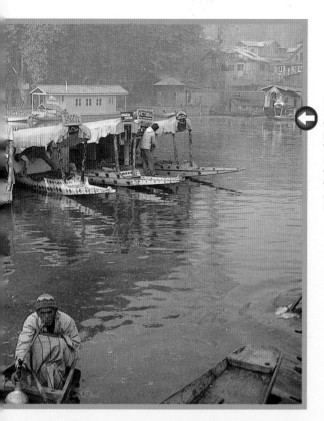

Shikaras, with canopied roofs and fluttering curtains, glide along one of Srinagar's many canals. The Jhelum River, spanned by seven wooden bridges, runs through this summer capital of Jammu and Kashmir.

The Vale of Kashmir is a welcome and beautiful sight as drivers emerge from the Jawahar Tunnel in the Banihal Pass. Built in the 1970's at a height of about 9,000 feet (2,750 meters), the tunnel allows year-round travel through the mountain pass.

Indian security forces patrol the fenced border with Pakistan in Suchetgarh, southwest of Jammu. Since 1947, India and Pakistan have been locked in a dispute over which country should govern Kashmir. Currently the region is controlled partly by India, partly by Pakistan, and partly by China.

A troubled paradise

Since 1947, Kashmir has been the center of a dispute between India and Pakistan. A truce line now divides Jammu and Kashmir into Indian and Pakistani sections. Neither country recognizes the jurisdiction of the other.

In the late 1980's, protests erupted in Indian-held Kashmir. Some protesters wanted Kashmir to join Pakistan, and others wanted independence for Kashmir. By 1990, the protests had developed into armed combat between guerrillas and Indian security forces. Fighting has killed many thousands of people.

In 1999, fighting broke out between India and Pakistan after Pakistani troops occupied the Kargil area on the Indian side of the truce line. A terrorist attack on India's Parliament building in 2001 led India to build up forces along the line. Pakistan responded in kind, and the two countries came to the brink of war. The following year, international diplomacy eased the tensions. Since then, India and Pakistan have attempted to work toward a peaceful solution to the Kashmir conflict.

INDONESIA

The Southeast Asian country of Indonesia consists of more than 17,500 islands that extend across more than 3,200 miles (5,150 kilometers) along the equator. Many of Indonesia's islands cover less than 1 square mile (2.6 square kilometers). However, Indonesia also includes about half of New Guinea and three-fourths of Borneo—the second and third largest islands in the world.

Tropical rain forests with many valuable hardwood trees cover much of Indonesia's land area. Crocodiles, elephants, pythons, rhinoceroses, and tigers live in some of these forests. Indonesia's mountains include many volcanoes, a number of which are still active. Although several violent volcanic eruptions have killed many people, Indonesians still live and farm near the volcanoes, where volcanic ash makes the soil extremely fertile.

Today, about half the Indonesian people live in small farm villages. Many of them still follow traditional ways of life.

Indonesia has an extremely long history. Scientists have found bones in Java of one of the earliest types of prehistoric human beings. Now known as *Java man,* this prehistoric human being may have lived in the area as far back as 1.8 million years ago.

As early as 2500 B.C., the ancestors of the Indonesian people established ocean trade routes between the islands. The islands became a crossroads for trading ships traveling between China and Arabia.

In early times, the region from India to Japan, including Indonesia, was known to Europeans as the Indies. When Christopher Columbus arrived in America, he was really looking for a westward sea route from Europe to the Indies. He hoped to find riches in the Indies and to establish a great city for trading the products of the East and the West.

By the 1500's and 1600's, Portugal, England, and the Netherlands struggled to gain control of Indonesian trade. The Dutch controlled the trade on most of the islands by the late 1700's, and they expanded political control over most of region during the 1800's. Japan occupied the islands from 1942 to 1945, during World War II. Following the war, Indonesian nationalists achieved independence.

Today, Indonesia has an extremely diverse population. Its people belong to about 300 different ethnic groups and speak more than 250 languages. This mixture has created a rich and varied culture, visible in many ways, including food and clothing, arts and architecture, and music and dance.

INDONESIA TODAY

Indonesia is a mix of old and new. About half of all Indonesians live in small, traditional farm villages, but the country also has bustling cities full of high-rise buildings. One such city is Jakarta, the capital, on the island of Java.

In 1908, while Indonesia was still a colony of the Netherlands, its first nationalist organization was founded. In 1927, a civil engineer named Sukarno founded the Indonesian Nationalist Party. In 1945, Sukarno and other nationalists declared Indonesia's independence. Sukarno became president of the Republic of Indonesia, and a constitution was soon established.

The Dutch tried to regain control of the nation and, from 1945 to 1949, the Dutch and Indonesians fought many bitter battles. Although the Dutch recaptured much territory, they were unable to defeat the Indonesians. In November 1949, under pressure from the United States and the United Nations (UN), the Dutch agreed to grant independence to all of Indonesia except the western part of the island New Guinea. Indonesia acquired this area during the 1960's.

Parliamentary elections, held in 1955, failed to produce a majority party. Revolts against the government broke out in 1958, but army units from Java defeated all the rebels by 1961.

Guided democracy

In 1960, Sukarno dissolved the elected parliament and appointed a new one. He called his system of government *guided democracy*. In 1963, Sukarno was declared president for life.

Under Sukarno, the country became almost bankrupt. In 1966, pressure from the army and student groups forced Sukarno to transfer much of his political power to General Suharto, the army leader. In 1968, Suharto became Indonesia's second president, and he established a government known as the New Order. Initially, Suharto improved political and economic stability.

In 1998, violent riots and a deepening economic crisis led to the ouster of Suharto. He resigned in May of that year. In October 1999, the People's Consultative Assembly elected a new president. Since 2004, the Indonesian people have directly elected the nation's president.

Indonesia, an island country in Asia, is about 20 percent as large as the continental United States. It stretches for 3,200 miles (5,150 kilometers) between the Indian and Pacific oceans.

FACTS

Official name:	Republik Indonesia (Republic of Indonesia)
Capital:	Jakarta
Terrain:	Mostly coastal lowlands; larger islands have interior mountains
Area:	735,358 mi² (1,904,569 km²)
Climate:	Tropical; hot, humid; more moderate in highlands
Main rivers:	Mahakam, Musi, Kayan
Highest elevation:	Puncak Jaya, 16,503 ft (5,030 m)
Lowest elevation:	Indian Ocean, sea level
Form of government:	Republic
Head of state:	President
Head of government:	President
Administrative areas:	33 provinsi-provinsi (provinces)
Legislature:	Dewan Perwakilan Rakyat (House of Representatives) with 550 members serving five-year terms
Court system:	Mahkamah Agung (Supreme Court)
Armed forces:	302,000 troops
National holiday:	Independence Day - August 17 (1945)
Estimated 2010 population:	239,781,000
Population density:	326 persons per mi² (126 per km²)

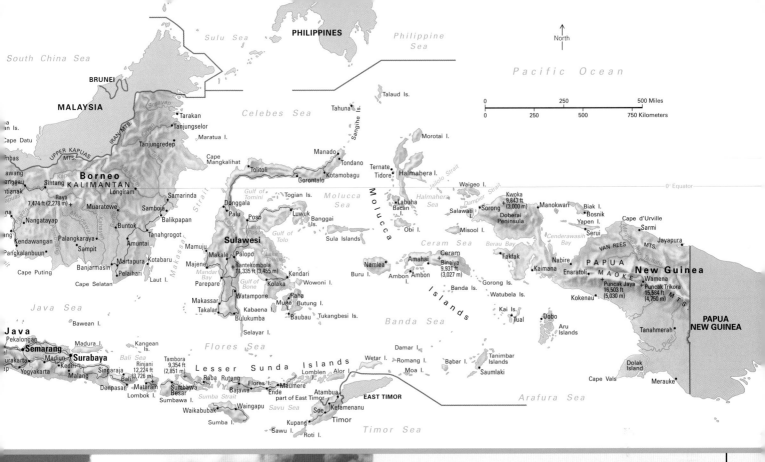

Population distribution:	51% rural, 49% urban
Life expectancy in years:	Male, 69; female, 73
Doctors per 1,000 people:	0.1
Birth rate per 1,000:	19
Death rate per 1,000:	6
Infant mortality:	30 deaths per 1,000 live births
Age structure:	0-14: 28%; 15-64: 66%; 65 and over: 6%
Internet users per 100 people:	11
Internet code:	.id
Languages spoken:	Bahasa Indonesia (official), English, Dutch, local dialects
Religions:	Muslim 86.1%, Protestant 5.7%, Roman Catholic 3%, Hindu 1.8%, other 3.4%
Currency:	Indonesian rupiah
Gross domestic product (GDP) in 2008:	$512.32 billion U.S.
Real annual growth rate (2008):	6.1%
GDP per capita (2008):	$2,206 U.S.
Goods exported:	Clothing, coffee, crude oil, electronics, natural gas, palm oil, rubber, textiles, wood and wood products
Goods imported:	Chemicals, crude oil and petroleum products, food, iron and steel, machinery, transportation and electrical equipment
Trading partners:	Australia, China, Japan, Malaysia, Singapore, South Korea, Thailand, United States

Struggles for independence

A number of separatist movements have arisen in parts of Indonesia. In 1975, Indonesia, which governed the western part of the island of Timor, invaded the eastern half of the island. Many people in East Timor resisted Indonesian rule. They voted for independence in a UN-sponsored vote in 1999 and finally got it in 2002.

In 2002, Indonesia also gave Irian Jaya (the name it had given the western part of New Guinea) greater control over local affairs and renamed the province Papua, which its people preferred. In December 2002, Indonesia's government reached a peace settlement with separatist Sumatran rebels in Aceh, but the settlement broke down in 2003. Talks resumed in 2005, and in 2006, Indonesia gave Aceh greater self-government.

Natural disasters

In December 2004, an undersea earthquake near Sumatra generated a tsunami that killed about 168,000 people in Indonesia and left millions of people homeless. In May 2006, a major earthquake struck the island of Java. More than 5,800 people were killed. In July, a tsunami hit the southern coast of Java, killing more than 600 people.

ENVIRONMENT

Indonesia has a hot, humid climate. The lowlands have an average temperature of about 80° F (27° C), but temperatures are lower in the highlands. Average temperatures vary little throughout the year. Java and the Lesser Sunda Islands are the only parts of the country with a dry season.

The islands of Indonesia can be divided into three groups: the Greater Sunda Islands, where most Indonesians live and where most economic activity is centered; the Lesser Sunda Islands, one of the few areas with a dry season; and the Moluccas, once known as the Spice Islands. Indonesia also includes Papua, which is part of the large island of New Guinea.

The Greater Sunda Islands

This island chain includes Borneo, Sulawesi, Java, and Sumatra. Of the four islands, Java has the most people and the largest city, Jakarta.

Borneo, the largest island in the Greater Sundas, has thick tropical rain forests and mountains in most of its interior. The southern part of Borneo—about three-fourths of the island—belongs to Indonesia. The area is thinly populated, and most people live along the coast. The Kapuas River, the longest river in Indonesia, flows about 700 miles (1,100 kilometers) from the mountains to the sea.

Forested Sulawesi (formerly Celebes), to the east of Borneo, is Indonesia's most mountainous island. Volcanoes, some of them active, rise on the northern peninsula. Some of Sulawesi's inland valleys and plateaus have fertile farmlands and rich grazing lands, while the coastal waters provide a bountiful catch.

Java, the most industrialized island of Indonesia, has most of Indonesia's large cities, as well as thousands of small farm villages. An east-west chain of mountains extending across Java includes many volcanoes, some of them active. Wide, fertile plains lie north of the mountains, with limestone ridges to the south. A large highland plateau covers western Java.

Sumatra, to the west of Java, is the sixth largest island in the world. Along the southwestern coast of the island, a range of volcanic peaks rises about 12,000 feet (3,660 meters). The mountains slope eastward to a broad plain covered by dense rain forests and some farmland. Much of the eastern coast is swampy.

The Lesser Sunda Islands

This island chain extends eastward from Bali about 700 miles (1,100 kilometers) to Timor, the largest of the Lesser Sundas. Among these islands, Bali has the largest population and the largest city, Denpasar.

The Lesser Sundas have many mountains, and many small rivers flow from the mountains to the sea. The eastern islands have fewer rain forests and more dry grasslands than those in the west. Corn is the main crop in the east, while rice ranks first in the west. Most towns are coastal trading centers.

Lake Toba lies in the north central region of Sumatra. It is among the world's largest calderas. A caldera is a deep depression that once formed the magma chamber of a volcano.

Farmers cultivate rice on small farms and on many kinds of land in Indonesia. In mountainous Bali, farmers build terraces and dikes (dirt walls) to provide more land for growing rice, the basic food of Indonesians.

Mount Bromo, in eastern Java, is one of a chain of volcanoes that crosses the major islands of Indonesia. Java has 112 volcanoes, and some are still active.

The Moluccas

The Moluccas lie between Sulawesi and New Guinea on both sides of the equator. Scattered among the large islands of the chain are hundreds of coral islands and *atolls,* groups of islands that enclose or partially enclose a lagoon. Most of these small islands are unpopulated. Almost all of the large islands of the Moluccas are mountainous and thickly forested.

The people of the Moluccas live primarily in coastal trading settlements. Ambon, an important port, is the largest city. The Moluccas became famous hundreds of years ago, when traders gathered spices there for sale in Asia, Africa, and Europe. They were known as the Spice Islands.

Papua

Papua is a province of Indonesia formerly known as Irian Jaya. It consists of the western half of the island of New Guinea and some small nearby islands. This area, the least developed and most thinly populated part of Indonesia, is a region of tropical rain forests and towering, snow-capped mountains. Puncak Jaya, the highest mountain in Indonesia, rises 16,503 feet (5,030 meters) over Papua. Cities, towns, and most farmland lie along the low, swampy coasts.

The volcano on the island of Krakatau, which lies between Sumatra and Java, erupted in 1883 (A), causing one of the world's worst disasters. Much of Krakatau Island disappeared. Volcanic ash rose 50 miles (80 kilometers), and a tidal wave killed about 36,000 people. Sections of the volcano that project above the water today form the islands Krakatau, Anak Krakatau, Lang, and Verlaten. (B)

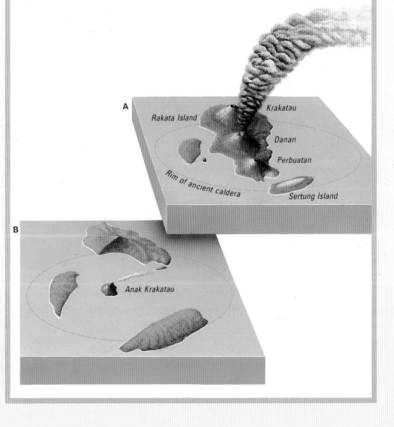

PEOPLE

Indonesians belong to about 300 different ethnic groups. About 40 percent of the people are Javanese, and about 15 percent are Sundanese. Other groups include Bantenese, Betawi, Bugis, Chinese, Madurese, Malays, and Minangkabau.

More than 250 languages are spoken in Indonesia, including Bahasa Indonesia, the country's official language. Most children learn the language of their region at home before starting school. Later, when they enter school, they learn Bahasa Indonesia.

As of 2010, more than 239,000,000 people lived in Indonesia. Java is the most densely populated island, with about 60 percent of the country's population. Because Java is so crowded, the government has encouraged people to move to less populated islands. However, Java's population continues to increase rapidly.

Many Indonesians, especially those born in Java, have only one name. For example, Sukarno, the country's first president, and Suharto, who was president from 1968 to 1998, had only one name.

Traditional life

Small farming villages are home to about half of Indonesia's people. Village headmen and other traditional leaders, such as religious teachers, control most of village life. These traditional leaders govern by a system of local customs that stress cooperation. The villagers often settle disputes and solve problems by holding an open discussion that continues until an agreement is reached.

Most rural Indonesian families live in houses that consist of a sleeping room and a large living room. The living room may also serve as the kitchen. Most traditional Indonesian houses stand on stilts, and families use the space underneath the houses for cattle stalls, chicken coops, or storage for tools and firewood.

Some Indonesian groups build *long houses* that may house up to several hundred people. These groups include the Dayaks in Borneo, the Batak in Sumatra, the Toraja or Sulawesi, and some Papuan groups in Papua. Many Indonesians decorate their walls with beautifully carved wood panels.

An Indonesian family rides on a motorcycle in Jakarta. Very few Indonesians own an automobile. Instead, they travel by bicycle, bus, motorcycle, train, and in six- to eight-seat vehicles called *bemos*.

A Toraja village on the island of Sulawesi is home to one of Indonesia's many small ethnic groups. Many of the Toraja houses and granaries are decorated with beautifully carved wood panels.

More than 85 percent of Indonesia's people are Muslims, about 9 percent are Christians, and about 2 percent are Hindus. However, many Indonesians still hold other traditional religious beliefs and combine worship of ancient ancestors and nature with other religions.

Traditional Indonesian clothing for both men and women is a *sarong*. A sarong is a colorful skirt that consists of a long strip of cloth wrapped around the body.

Many Indonesian dishes are flavored with the spices that once made the islands famous. Rice, the country's chief food crop, is eaten with meat, fish, vegetables, or hot spices. Indonesian food is often cooked in coconut milk and oil and served on banana or coconut leaves.

Modern life

Less than 10 percent of Indonesia's people could read and write in 1945, when the government launched literacy programs conducted mainly in the villages. Today, most Indonesians can read and write. Indonesian law requires children to attend elementary school for nine years, beginning at least by age 7. Currently, nearly all of the country's children attend school. Indonesia has about 50 universities.

Recreation in Indonesia has both traditional and modern influences. Indonesians in cattle-breeding areas hold ox races and bullfights. In some areas, people enjoy *Pencak silat,* an activity that combines dancing with self-defense.

Indonesians also enjoy several Western sports, including badminton, basketball, and soccer. Many sports events are held in Jakarta's 200,000-seat stadium.

Jakarta, Indonesia's capital, is the nation's most modern city. High-rise hotels and modern office and government buildings dominate the city's center, while major industries operate on the outskirts. Cars, buses, taxis, trucks, and motorcycles, as well as an electric railroad and an international airport, serve the people of Jakarta. The capital also has several universities, including the University of Indonesia.

Jakarta has several fashionable residential areas south of the city. Many of its people, however, live in small wood or bamboo structures in areas called *kampongs*. Often, the kampongs are without clean water, sewers, or electricity.

Indonesia's population is uneven and widely distributed. People live on 6,000 of the islands, but the islands in the center of the country are most heavily populated. The most thinly populated region is Papua, the western half of New Guinea.

Persons per mi²	Persons per km²
More than 250	More than 100
125 to 250	50 to 100
60 to 125	25 to 50
25 to 60	10 to 25
Less than 25	Less than 10

Major urban centers
- More than 2.5 million inhabitants
- 1 million to 2.5 million inhabitants
- 200,000 to 1 million inhabitants

ART AND CULTURE

The Indonesian islands were a crossroads in the trade that extended from Arabia to China. Merchants of many lands, including Arabs, Chinese, Indians, and Persians, traded such goods as porcelain, textiles, and raw silk for Indonesian spices and sweet-smelling woods. Asian goods were also carried to Europe. Marco Polo, an Italian traveler, visited the islands in 1292.

The cultural contact that accompanied trade with the people of China, India, and the Arab world left an indelible mark on the Indonesian islands. The Buddhist, Hindu, and Islamic cultures in particular had a lasting effect on the religion, customs, and arts of the area.

Early influences

The ancient inhabitants of Indonesia established trade among the islands and with merchants on the Asian mainland as early as 4,500 years ago.

Beginning in the A.D. 400's, Hinduism and Buddhism began to have a strong influence in Indonesia. Indian architecture influenced the design of Indonesian temples, and Indian legends became part of local puppet plays in the villages. Although villagers continued to worship according to their ancient beliefs, they also began to pray to Indian gods. Hindu and Buddhist kingdoms battled for power in Indonesia for hundreds of years. Many ancient temples from both religions remain standing, but few Indonesians today are Buddhists or Hindus.

The Muslim impact

Traders from Arabia and India were among the first people to bring the faith of Islam to Indonesia, but the greatest influence came from events in Melaka. A port kingdom on the Malay Peninsula, Melaka became a major trading power during the 1400's. When Melaka's ruler converted to Islam, the religion began to spread throughout the islands.

Today, the vast majority of the Indonesian people are Muslims, but many are less strict in their observance of Islamic teachings than are Muslims in Arab countries. Indonesian Muslims have their own understanding of the religion, combining the beliefs of Islam with their own traditional culture.

Traditional ceremonial marches are popular in Indonesia. Every member of the family helps carry the elegant banners.

Orchestras called gamelans play an important part in Indonesian culture. The many gongs and drums of the gamelan produce a highly rhythmic sound. This music accompanies dance and shadow-puppet performances, sometimes lasting all night.

An Indonesian dancer is made up to look like a fierce animal. Many regional dances imitate the movements and habits of animals.

The legong, a Balinese folk dance, forms part of the anniversary rites of a village temple. The popular dance, usually performed by two or three girls, tells an ancient story of love and battle.

Literature, dance, drama, and crafts

Indonesia's arts include elements of the Hindu, Buddhist, and Islamic cultures. Ancient Hindu stories, for example, provide the basis for many of Bali's dramatic folk dances. The dancers wear elaborate costumes and headdresses and move to forceful rhythms.

Shadow-puppet dramas are also popular in Java and Bali. In these performances, which often last from 9 p.m. until 6 a.m., the shadows of the puppets are projected onto a white screen by a palm-oil lamp. Puppet plays based on Hindu myths are the most popular form of theater in Indonesia. An orchestra called a *gamelan* accompanies the dances and puppet plays. The gamelan has a distinctive sound created by metal gongs, flutes, xylophone-like instruments, double-ended drums, and a stringed instrument played like a cello.

In Indonesia's literature and crafts as well, regional cultures mix with the Buddhist, Hindu, and Islamic. Early Indonesian literature, for example, consisted largely of local folk tales and traditional Hindu and Islamic stories. Modern literature often addresses the relationship between traditional Indonesian values and the modern world. Indonesian sculptors created beautiful decorations for their many ancient Hindu and Buddhist temples. The Balinese still carve Hindu figures and symbols for their homes and temples.

ECONOMY

Indonesia is a fertile land, rich in natural resources. These resources serve as the basis of the nation's economy.

About two-fifths of Indonesia's people work in agriculture and fishing. The country's many volcanoes have been good for its agriculture. The volcanic ash in the soil is rich in nutrients. Farmers raise such crops as cocoa, coffee, palm oil, rice, rubber, sugar cane, spices, tea, and tobacco for export. Small farmers also produce many food crops for local use, such as bananas, cassava, coconuts, corn, peanuts, soybeans, spices, and sweet potatoes. Some farmers also raise water buffalo, cattle, goats, hogs, sheep, and poultry.

Fish is a major part of the Indonesian diet. The country also exports seafood, especially shrimp and tuna.

Rice and rubber

Rice is the chief food crop of Indonesia. It is boiled or fried in a variety of ways and served with many other foods.

No one knows exactly when or where rice originated, but it probably grew wild and was gathered and eaten by people in Southeast Asia thousands of years ago. Today, Indonesia is one of the world's top rice-growing countries. Only China and India produce more rice.

Farmers in Java grow most of Indonesia's rice, producing at least two rice crops a year. Banks of earth are piled around the flat land to form fields called *paddies,* which are irrigated by water from mountain streams. In hilly areas, farmers plant rice on terraces built into the hillsides.

Indonesia is also one of the top producers of natural rubber. Almost all natural rubber comes from huge rubber tree plantations. Rubber trees grow best in hot, moist climates in acidic, well-drained soils. The world's finest rubber-growing regions lie within a *rubber belt* that extends about 700 miles (1,100 kilometers) on each side of the equator.

Coconut palms are grown on small farms all over Indonesia. The people use coconut as a flavoring in cooking and often serve food on coconut palm leaves. They also use the leaves to weave baskets and to thatch roofs.

Petroleum and timber

Petroleum and natural gas are Indonesia's chief exports. Indonesia is one of the chief producers of petroleum in the Far East. Most of the country's petroleum comes from East Kalimantan and Sumatra.

Indonesia is also a leading producer of copper and tin. The islands of Bangka and Belitung have many tin mines, and Papua produces most of the copper.

Wood products are major exports, but Indonesia has had to pass laws to prevent exploitation of its valuable forests. Timber comes mainly from Borneo and Sumatra. Poor inland transportation on most of the other large islands of Indonesia interferes with the development of lumbering there. Indonesia's valuable hardwoods include ebony and teak, which are known for their beautiful grain patterns and are widely used in cabinets, flooring, furniture, and paneling. Cinchona bark is harvested to make *quinine,* a drug used to treat malaria.

A Balinese farmer takes his ducks to market. Raising ducks, which requires little space, is a good way to make a living on the crowded island of Bali.

Indonesia's paddies produce the population's main crop and staple food. Indonesian rice is grown mostly on small farms in irrigated fields, which are drained when the grain begins to ripen

Tropical resorts attract tourists from throughout the world. Visitors also flock to the island of Bali, which is renowned for its dancing and colorful festivals, and to the island of Java to enjoy its beautiful scenery and famous temples.

Manufacturing and service industries

Java is the most industrialized island. Its main industrial centers are Jakarta and Surabaya, and Indonesia's main seaport is Tanjung Priok, near Jakarta. More than half of Indonesia's exports come from manufactured products. Indonesia's wide range of industries includes the manufacture of cement, chemicals, garments and footwear, plywood and other wood products, and textiles. Other important manufactured goods include cigarettes, electronic equipment, fertilizers, pharmaceuticals, processed rubber products, pulp and paper, and steel products.

Service industries, which provide services rather than produce goods, are becoming increasingly important. The main service industries include banking, government, trade, tourism, and transportation.

In 1997 and 1998, Indonesia suffered one of the worst financial slumps in its history. The value of its currency fell, and its stock market plunged. Banks and other businesses failed, and millions of people lost their jobs. Meanwhile, the price of food and other necessities soared. The crisis led to violence and forced President Suharto to resign. The economy gradually recovered in the early 2000's, as the government introduced reforms in the financial sector and worked to stimulate economic growth.

RAIN FOREST UNDER THREAT

A tropical rain forest is a dense forest of tall trees in a region of year-round warmth and plentiful rainfall. Almost all such forests lie near the equator. Significant areas of both Malaysia and Indonesia are covered by these great forests.

About half the world's species of plants and animals live in tropical rain forests, even though rain forests cover only about 6 to 7 percent of the planet's surface. A rain forest has more kinds of trees than any other area in the world. More species of amphibians, birds, insects, mammals, and reptiles live in these forests than anywhere else on Earth.

The loss of the rain forest

Today, the rapid growth of the world's population and the increasing demands for natural resources threaten many tropical rain forests. People have destroyed large rain forests by clearing land for farms and cities. Mining, ranching, and timber projects have also caused great damage. This destruction of forests is called *deforestation*.

In 1950, tropical rain forests covered about 8.7 million square miles (22.5 million square kilometers) of Earth's land. By the early 2000's, only about half that amount was left. Many rain forest species cannot adapt to these rapid changes. Scientists estimate that the clearing of forests wipes out about 7,500 species per year.

Deforestation has many far-reaching effects. The trees and other plants in forests help preserve the balance of gases in the atmosphere through a process called *photosynthesis*. In photosynthesis, plants take in sunlight, carbon dioxide, and water, and they give off oxygen. Renewal of the oxygen supply is vital to the continuing survival of oxygen-breathing organisms, such as animals and people. The absorption of carbon dioxide by plants is also important. An increase in carbon dioxide in the atmosphere could severely alter Earth's climate.

Forests also serve the planet by soaking up large amounts of rainfall. They prevent the rapid runoff of water, which may cause erosion and flooding. The trees and other plants in forests also provide habitats for many living creatures. Scientists fear that further deforestation will lead to the extinction of hundreds of thousands of species of plants and animals.

The conservation predicament

Efforts at conservation, which involves the protection of the world's natural resources, face serious challenges today. One of the most difficult challenges is the need to find a compromise between protecting the environment and increasing—or even maintaining—agricultural and industrial production. Many Asian countries, for example, are hard-pressed to conserve their natural resources because the land must support so many people.

	Rainforest
	Other forest areas
	Deforested areas

South China Sea

PHILIPPINES

MALAYSIA

Medan

MALAYSIA

Celebes Sea

KALIMANTAN

HALMAHERA

Equator

PAPUA NEW GUINEA

SUMATRA

SULAWESI

CERAM

PAPUA

Java Sea

Jakarta

Banda Sea

Surabaya

JAVA

Arafura Sea

EAST TIMOR

Indian Ocean

TIMOR

AUSTRALIA

Indonesia's tropical rain forests once covered most of the islands. However, the agricultural and industrial development of the 1900's has threatened to exhaust this vast resource. Rain forests around the world face a similar threat.

Lush rain forests contain an incredible variety of plant and animal life. The tallest trees of a rain forest may reach a height of more than 165 feet (50 meters). A tropical rain forest is always green.

Southeast Asia has an average of 375 persons per square mile (145 per square kilometer)—about three times the world average population. Many of the region's forests have been cut down to produce timber and to clear land for farms and industries. In order to provide living and working space for people, it has been necessary to reduce the living space of wildlife.

Because every species plays a role in maintaining the balanced living systems of Earth, the loss of any species can threaten the survival of all life—including human beings. The goals of conservation can only be achieved through the combined efforts of the world's people.

Deforestation, the destruction of forests, endangers the hundreds of species of plants and animals that live in Southeast Asian rain forests. Many of Southeast Asia's people clear forests to earn a living as farmers or timber workers.

The siamang, the largest of the gibbons, lives in the forests of Indonesia and Malaysia. The number of apes is decreasing because human activity has destroyed much of their habitat.

IRAN

Iran is located in southwest Asia, north of the Persian Gulf. It lies east of Iraq and Turkey and west of Afghanistan and Pakistan. It is south of Armenia, Azerbaijan, Turkmenistan, and the Caspian Sea.

Iran is one of the oldest countries on Earth. Its history dates back almost 5,000 years to the first settlements of the Elamites in the southwestern region of present-day Iran. Later, the great Persian Empire included what is now Iran, as well as most of southwest Asia and parts of Europe and Africa.

Throughout its history, Iran has been the site of many invasions and conquests. One of the most important was the Arab invasion of the A.D. 600's. During this period, the Arabs converted the Persians to the religion of Islam. Today, most Iranians belong to the Shiah branch of Islam.

The discovery of oil in southwestern Iran during the early 1900's was another important development in Iran's history. Profits from oil exports promised to bring great wealth to the country. Reza Shah Pahlavi, who ruled Iran from 1925 to 1941, used these profits to modernize the nation and promote economic and social development.

His son, Mohammad Reza Pahlavi, became *shah* (king) in 1941 and introduced more economic and social reforms. The shah also used Iran's increasing oil revenues to develop many new industrial projects and provide a base for economic growth.

Many traditional Muslims disagreed with the shah's reforms, claiming that the actions violated Islamic teachings. In the late 1970's, opponents of the shah, led by the religious leader Ayatollah Ruhollah Khomeini, took control of the government and declared Iran an Islamic republic. Islamic tradition replaced many of the Western ways introduced by the shah. Khomeini led the government until his death in 1989. After Khomeini's death, some liberal reformers sought to lessen the government's control over people's personal lives and freedoms.

For both reformers and more conservative leaders, Islam and its teachings are extremely important. Friday prayers at Tehran University are an example of the nation's devotion to Islamic customs.

175

IRAN TODAY

The Islamic revolution of 1979 brought many changes to Iran. The government eliminated many Western influences and stopped almost all modernization. These political changes, along with the war with Iraq during the 1980's, weakened the country's economy.

Today, Iran faces many political and economic problems. Reformers want to change some of the conservative policies instituted in the 1980's and reduce government influence over personal lives. However, for the most part, conservative leaders have held on to power.

Government

Since Iran's new constitution was adopted in 1979, the country's supreme leader is the *faqih,* a scholar in Islamic law and the recognized religious leader of most Iranians. Ayatollah Khomeini was Iran's first faqih.

Iran's lawmaking body, the Majlis, is made up of 290 members elected to four-year terms. A Council of Guardians, consisting of six lawyers and six judges, examines all proposed laws to make sure they do not violate the Constitution or Islamic principles.

Way of life

Islamic tradition is the basis of life in Iran. The Iranian government requires schools to teach Islamic principles. It requires women to wear a *chador*—a full-length body veil worn over a woman's other clothes—or some other head covering, in order to comply with its interpretation of Islamic traditions about modesty.

All entertainment thought to be against the teachings of Islam is banned by the government. Freedom of speech and other civil rights have also been severely restricted.

After the revolution

The new government was bitterly anti-American because the United States had supported the shah. In October 1979, U.S. President Jimmy Carter allowed the shah to enter the United States for medical treatment. On November 4, Iranian revolutionaries seized the U.S. Embassy in Tehran. They held hostages for nearly two years while demanding the return of the shah for trial.

FACTS

Official name:	Jomhuri-ye Eslami-ye Iran (Islamic Republic of Iran)
Capital:	Tehran
Terrain:	Rugged, mountainous rim; high, central basin with deserts, mountains; small, discontinuous plains along both coasts
Area:	636,372 mi² (1,648,195 km²)
Climate:	Mostly arid or semiarid, subtropical along Caspian coast
Main rivers:	Karkheh, Zahreh, Mand, Atrek
Highest elevation:	Mount Damavand, 18,386 ft (5,604 m)
Lowest elevation:	Caspian Sea, 92 ft (28 m) below sea level
Form of government:	Islamic republic
Head of state:	Supreme leader
Head of government:	President
Administrative areas:	30 ostanha (provinces)
Legislature:	Majlis (Islamic Consultative Assembly) with 290 members serving four-year terms
Court system:	Supreme Court
Armed forces:	523,000 troops
National holiday:	Republic Day - April 1 (1979)
Estimated 2010 population:	74,131,000
Population density:	116 persons per mi² (45 per km²)
Population distribution:	67% urban, 33% rural
Life expectancy in years:	Male, 69; female, 72
Doctors per 1,000 people:	0.9
Birth rate per 1,000:	18
Death rate per 1,000:	6
Infant mortality:	32 deaths per 1,000 live births
Age structure:	0-14: 26%; 15-64: 69%; 65 and over: 5%
Internet users per 100 people:	32
Internet code:	.ir
Languages spoken:	Persian Farsi and Persian dialects, Turkic and Turkic dialects, Kurdish, Luri, Balochi, Arabic, Turkish
Religions:	Shiah Muslim 89%; Sunni Muslim 9%, others (Zoroastrian, Jewish, Christian, and Bahá'í) 2%
Currency:	Iranian rial
Gross domestic product (GDP) in 2008:	$382.30 billion U.S.
Real annual growth rate (2008):	6.5%
GDP per capita (2008):	$5,306 U.S.
Goods exported:	Mostly: crude oil Also: carpets, chemicals, fruits and nuts, iron and steel
Goods imported:	Food, iron and steel, machinery, motor vehicles, petroleum products
Trading partners:	China, Germany, Italy, Japan, South Korea, Turkey, United Arab Emirates

Iran has been an Islamic republic since 1979, when a revolution ended the Persian monarchy. The country lies north of the Persian Gulf. Much of Iran is covered by a barren wasteland of deserts and mountains.

In 1980, Iran began fighting a war with Iraq. Hundreds of thousands of Iranians were killed or injured, and over a million people were left homeless. After the cease-fire in 1988, Iran began trying to rebuild itself.

Ayatollah Khomeini died in 1989. Iran's top religious leaders chose President Ali Khamenei to succeed Khomeini as faqih.

During the 1990's, Iran's oil exports declined because of a decrease in production capacity and lower demands in the world market. Political discontent within Iran increased after Khomeini's death. Serious conflicts emerged among Iran's leaders, and growing numbers of people openly blamed government leaders for the state of the economy.

In the early 2000's, Iran's nuclear policy alienated many nations. Iran claimed its goal was to develop electrical power, but other nations feared the program could be used to develop materials for nuclear weapons. From 2006 to 2010, the United Nations imposed four rounds of sanctions on Iran in an attempt to prevent Iran from acquiring the materials necessary to build nuclear weapons.

In parliamentary elections held in 2000, members of reform groups won a majority of seats in the Majlis. These groups supported a number of liberal measures, including freedom of the press and less government influence over Iranians' personal lives. Conservatives regained control of the Majlis in 2004. They defeated reformists again in 2009 in elections that many Iranians considered fraudulent. Violence erupted as the government defended the validity of the elections.

Antigovernment protests erupted again in 2011 in cities throughout Iran. Protesters voiced support for democracy movements in Tunisia and Egypt. Government security forces broke up the demonstrations and arrested many of the protesters.

PEOPLE

Iran stands at one of the world's major crossroads, between the Near East and central Asia. Because of its location, people have migrated across the region for centuries. It has been the site of many civilizations and empires.

Ancient settlers

The first major civilization in what is now Iran was that of the Elamites. They may have settled in southwestern Iran as early as 3000 B.C.

During the 1500's B.C., groups of Aryans began to migrate into Iran from central Asia. Today, about two-thirds of the Iranian population are descended from the Aryans. The name *Iran* itself comes from a Persian word meaning *Land of the Aryans*.

The modern Persian language, called Farsi, developed from the language of the Aryans. When the Arabs conquered Iran in the A.D. 600's, Farsi absorbed many Arabic words and Arabic script.

Ethnic groups

The Persians, Iran's largest ethnic group of Aryan origin, form about 60 percent of the country's total population. The majority of Persians live in central Iran and on the slopes of the Elburz and Zagros mountains.

Other groups believed to be descended from the Aryans include the Gilanis and Mazandaranis of the north, the Kurds of the northwest, the Lurs and Bakhtiaris of the west, and the Balochis of the southeast.

Non-Aryan ethnic groups include the Azerbaijanis, the Khamseh, the Qashqais, the Turkomans, and Arabs. The Azerbaijanis, who speak a Turkic language, are the largest of these groups.

Since the revolution in 1979, the Islamic government has monitored and strictly controlled many aspects of social life. In response, a number of ethnic groups have protested for greater political and cultural independence. Fighting has broken out a number of times between government troops and members of such groups as the Balochis, Kurds, and Turkomans.

The Metro subway transports riders in Tehran. Public transportation is an important mode of transportation in Tehran. In rural areas, people generally travel by bicycle, donkey, horse, and mule.

Iranian women pray at the Eid al-Fitr prayer service in Tehran. The holiday marks the end of the Muslim holy fasting month of Ramadan. Almost all Iranians are Muslims.

A Bakhtiari nomad herds his goats and sheep next to electrical power lines near the town of Masjid-e-Soleiman in southwestern Iran. Migration has been a way of life for nomads for thousands of years but is slowly disappearing in modern times.

Religion

About 98 percent of the Iranian people are Muslims. About 90 percent of them belong to the Shiah branch of Islam, which is Iran's state religion. Most members of the country's largest ethnic groups, the Persians and Azerbaijanis, are Shiites.

Although the Islamic government has little tolerance for other religions, a few religious minorities do exist. Most of the Muslims who are not Shiites belong to the Sunni branch of Islam. Iran also has small numbers of Christians, Jews, and followers of Zoroastrianism, an ancient Persian religion.

The largest non-Muslim religious minority are the Bahá'ís. The Bahá'í faith grew out of the Bábi faith, a religion founded in Iran in 1844. Bahá'ís, who number about 250,000, are forbidden by the government to practice their faith. They have been severely persecuted for their beliefs.

Iranian refugees

About a million Iranians left their country after the Islamic revolution and the start of the war with Iraq. Some found shelter in neighboring countries, such as Turkey. Many others fled to Western Europe, Canada, and the United States.

Most of these refugees were middle-class Iranians who lived in the cities. They included technicians and scientists, as well as artists and teachers. The departure of these educated people left Iran in great need of skilled professionals.

LAND AND ECONOMY

Iran lies in an arid zone dominated by deserts and mountains. There are no major river systems. For centuries, people traveled across the country by caravan, following routes through gaps and passes in the mountains. The rugged landscape still makes transportation difficult and expensive today.

Landscape and climate

Iran's landscape can be divided into four regions: (1) the Mountains, (2) the Interior Plateau, (3) the Caspian Sea Coast, and (4) the Khuzistan Plain.

Iran has two mountain ranges, the Elburz and the Zagros. The Elburz Mountains stand along the northern border between the Caspian Sea Coast and the Interior Plateau. The northern slopes of the Elburz receive a great deal of rainfall, so the farmers are able to grow many crops in the region's rich soil. The southern slopes are barren and dry.

The Zagros Mountains begin at the borders of Turkey and Azerbaijan and stretch south and east to the Persian Gulf. The valleys of the northern and central part of the range are well populated. Few people live in the dry, rugged southern sections.

Winters are long and bitterly cold in the mountain regions. The summers are mild.

The Elburz and Zagros mountains surround the Interior Plateau of central and eastern Iran. The Interior Plateau stands about 3,000 feet (900 meters) above sea level and covers about half of Iran's area.

Most of the Interior Plateau consists of two deserts, the Dasht-e Kavir and the Dasht-e Lut. These deserts are among the most arid and barren on Earth. Together, they cover more than 38,000 square miles (98,000 square kilometers). Summer temperatures in the desert soar as high as 130° F (54° C). Rainfall is sparse—only about 2 inches (5 centimeters) per year. Temperatures in the cities of the plateau are less extreme.

The Caspian Sea Coast is a narrow strip of lowland between the Caspian Sea and the slopes of the Elburz Mountains. Because of its mild climate and plentiful rainfall, the Caspian Sea Coast is the most heavily populated region in Iran.

The Khuzistan Plain lies north of the Persian Gulf, between the border of Iraq and the Zagros Mountains. It is a very important region for Iran, because it contains the country's richest oil deposits. Many crops are also grown in the Khuzistan Plain's rich farmland. Summers on the plain are very hot and humid, but winters are mild and pleasant.

The economy

Iran's government controls most of the country's economy. Iran is a major petroleum producer. Oil dominates the economy, even though mining employs less than 1 percent of the country's workers. Iran is also a leading producer of natural gas. Other mining products include aluminum, cement, coal, copper, iron ore, lead, and zinc.

Women weave a carpet in a farmyard near Persepolis in southwestern Iran. Persian rugs are still made by artisans in private homes.

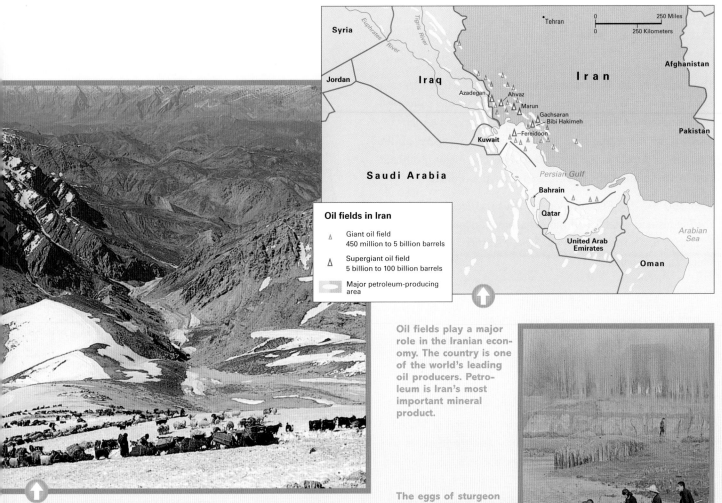

Oil fields in Iran

△ Giant oil field
450 million to 5 billion barrels

△ Supergiant oil field
5 billion to 100 billion barrels

▬ Major petroleum-producing area

Oil fields play a major role in the Iranian economy. The country is one of the world's leading oil producers. Petroleum is Iran's most important mineral product.

The eggs of sturgeon caught in the Caspian Sea make a delicacy called caviar. Sturgeon eggs are the most important product of Iran's fishing industry.

Bakhtiari herders drive flocks of sheep and goats on their yearly migration over the Zagros Mountains. These nomads live in round tents made of black felt.

Unwanted natural gas is burned off at an oil field on Kharg Island in the Persian Gulf off the Iranian coast. Natural gas remains a significant mining product in Iran's economy.

Service industries provide about half the jobs in the country. The main service industry employers are government agencies, hospitals, schools, banks, insurance companies, and community services.

Manufacturing is also an important part of the economy. The chief manufactured products are food products, petroleum products, and petrochemicals.

Agriculture employs about a fourth of all Iranian workers, yet the country still must import a large amount of food, because so much of the land is too dry to farm. The chief agricultural products include barley, corn, cotton, dates and other fruits, nuts, rice, sugar beets, and wheat. People also raise cattle, chickens, and sheep.

HISTORY

The history of Iran begins as far back as 3000 B.C., when an ancient people called the Elamites lived in what is now southwestern Iran. About 1,500 years later, groups of Aryans began moving into the region from central Asia. One group set up the kingdom of Media in the northwest. A second major group settled in the south of the region, later known as Persia.

In 550 B.C., the Persians, led by Cyrus the Great, overthrew the Medes and established the Achaemenid dynasty. By 539 B.C., they had added Babylonia, Syria, Palestine, and all of Asia Minor to their empire.

The golden age of Persia

In 522 B.C., Darius I became king of the Achaemenid Empire. The empire prospered under his reign and stretched from what is now Pakistan in the east to Libya in Africa and land west of the Black Sea in Europe.

By the mid-400's B.C., the glory of the Achaemenid Empire was almost gone. In 331 B.C., Alexander the Great of Macedonia conquered the empire. After Alexander's death, one of his generals, Seleucus, founded a new dynasty, the Seleucid. The Seleucids governed Iran until about 250 B.C., when the Parthians defeated them. In A.D. 224, the Persians overthrew the Parthians and established the Sasanian dynasty.

The rise of Islam

The Arabs conquered the Sasanians in the mid-600's and gradually converted most Iranians to Islam. Arabic replaced Persian as the official language of government in Iran, but most common people continued to speak Persian. Under the reign of the Arabs, Iran became a world center of art, literature, and science.

By the mid-1000's, Seljuk Turks had conquered most of Iran. In 1220, the Mongols, led by Genghis Khan, conquered Iran, destroying many cities and killing thousands of people.

In the late 1400's and early 1500's, the Safavids, a family of Persian descent, gained control over several regions of Iran. The Safavid kings ruled Iran until 1722, when armies from Afghanistan invaded.

Nadir Shah, a Turkish tribesman, drove the Afghans out of Iran in the 1730's and became king. After Nadir Shah was assassinated in 1747, civil war broke out in Iran, as many leaders fought for control of the country.

TIMELINE

c. 3000 B.C.	The region that is now Iran is settled by Elamites.
c. 1500's B.C.	Aryan settlers begin to arrive in Iran
550 B.C.	Cyrus the Great overthrows Media and founds Achaemenid Empire.
539 B.C.	Cyrus completes the conquest of Babylonia, Palestine, Syria, and Asia Minor.
331 B.C.	Alexander the Great conquers Persia.
323-250 B.C.	The Seleucids rule Iran.
250 B.C.	Parthian armies conquer Iran.
A.D. 224	Persians overthrow Parthians and found Sasanian dynasty.
mid-600's	Muslim Arabs conquer Iran.
Mid 1000's-1220	Seljuk Turks rule Iran.
1220	Mongols led by Genghis Khan conquer Iran.
1501-1722	The Safavid dynasty rules Iran.
1736	Nadir Shah conquers Iran.
1794-1925	The Qajar dynasty rules Iran.
1826	Russia invades Iran.
1828	Treaty of Turkomanchai establishes boundaries with Russia.
1856-1857	Iran invades Afghanistan. Iran signs peace treaty with United Kingdom, giving up claims to Afghan territory.
Early 1900's	Anglo-Persian Oil Company develops oil fields in Iran.
1906	Shah Muzaffar al-Din signs Iran's first constitution.
1925	Reza Khan (Reza Shah Pahlavi) becomes shah.
1941	Mohammad Reza Pahlavi succeeds his father to the throne.
1951	Iran's oil industry is nationalized.
1979	Revolutionaries force shah out of Iran and establish Islamic republic.
1980	Outbreak of war with Iraq.
1987	Dispute with United States in Persian Gulf over safety of Kuwaiti oil tankers.
1988	Iran-Iraq War ends in cease-fire.
1989	Death of Ayatollah Khomeini, leader of the 1979 revolution and faqih of Iran.
Early 2000's	Iran's nuclear policy leads to tensions with the United States and other nations.

In the late 1700's the Qajar dynasty came to power and ruled Iran until 1925. Reza Khan became shah in 1925 after overthrowing the Qajar government. He then changed his family name to Pahlavi. Reza Shah tried to modernize Iran and eliminate foreign interference.

Soon after World War II began in 1939, Iran declared itself neutral. However, the Allies wanted to use the Trans-Iranian Railway to ship war supplies to the Soviet Union. Reza Shah objected, and in 1941, Soviet and British troops invaded Iran. They forced Reza Shah to give up his throne. His son, Mohammad Reza Pahlavi, became shah. The new shah signed a treaty with the Allies, allowing them to ship supplies through Iran.

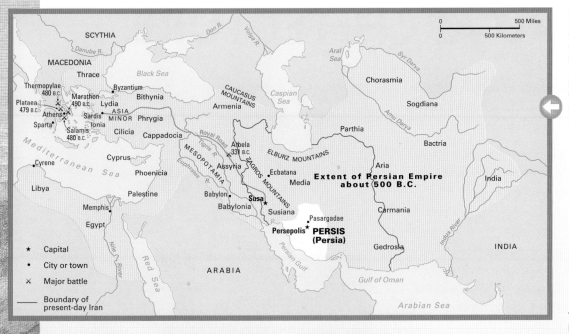

During the 500's B.C., the Persian Empire extended from India in the East to the Mediterranean Sea in the West. Cyrus the Great established the empire, which further prospered a few years after Cyrus's death under the reign of King Darius I.

The royal mosque at Isfahan, built in 1628-1629, shows the influence of many different cultures on Iranian architecture.

Khosrau II (died A.D. 628)

Mohammad Reza Pahlavi (1919-1980).

Ayatollah Ruhollah Khomeini (1900?-1989)

During the early 1960's, the shah began a series of economic and social reforms. Like his father, Mohammad Reza Pahlavi wanted to modernize Iran.

The shah ruled with absolute control over the government. Many Iranians opposed his use of the SAVAK—a secret police force. Many conservative Iranians also believed that his modernization programs violated traditional Islamic teachings.

In January 1979, the shah was forced to leave the country. The following month, revolutionaries led by Ayatollah Ruhollah Khomeini took over the government and established Iran as an Islamic republic.

A new constitution established the office of *faqih*, a religious scholar who is the supreme leader of the country. Khomeini was faqih until his death in 1989. He was succeeded by Ali Khamenei, who is called the *Spiritual Guide* or *Supreme Guide* of the nation.

ART AND ARCHITECTURE

Iran's artistic heritage begins with its earliest civilizations. Painted pottery dating back to about 5000-4000 B.C. has been found in western Iran.

From these beginnings, Iranian artists and architects have given the world some of its most splendid paintings, architecture, metalwork, ceramics, and textiles.

The Achaemenid Empire

When Cyrus the Great united Persia and Media under his rule in 550 B.C., he established the great Achaemenid Empire. The Achaemenids expressed the glories of their rule through art and architecture.

The Achaemenid palaces were huge, impressive structures. The palace at Persepolis, the capital of the Achaemenid Empire, features a large number of rooms, halls, and courts set on a raised platform. The structure blends the influence of many different cultures to create a uniquely Persian style. Its columns may have been 60 feet (18 meters) high. The columns show the influence of Egyptian architecture, and the ornamental detail at their bases came from the Ionian Greeks in Asia Minor.

Sasanian art and architecture

The Sasanian Empire was known for its metalwork. Goblets, plates, and other objects were made of gold and featured fine ornamental detail. Sasanian artists also carved enormous sculptures in rock.

The Sasanian Empire's chief glory, however, was its woven silk. Large amounts of Sasanian woven silk were exported to Constantinople and the Christian West. Its many colors and patterns influenced the art of the Middle Ages, as well as Islamic art.

Islamic art

When the Arabs invaded Persia in the A.D. 600's, they brought with them the religion of Islam. The people who accepted Islam blended the influence of the highly developed art of Persia with the religious customs of their new-found faith. They developed a uniform style of art known as *Islamic art*.

A Persian rug from the 1500's shows the typical dark colors and floral patterns. Islamic craft workers developed carpet weaving into a fine art.

A painting dating from the 1300's portrays a scene from Persia's most famous epic poem, the *Shah-Nama (Book of Kings)*, written by Firdausi about A.D. 1010.

Columns that once supported a wooden ceiling over the Audience Hall of Darius I around 500 B.C. still stand at the site of ancient Persepolis.

Islamic art is noted for its mastery of technique, design, and color. Islamic artists sought to make all aspects of life beautiful. Examples of Islamic art can be seen in architecture, textiles, metalware, pottery, carved and molded plaster, glassware, wood and ivory carvings, and book *illuminations* (decorations). Islamic art flourished from the mid-700's to the 1700's.

Islamic teachings forbid Muslims to create images of living things. Muslims believe that *Allah* (God) is the one and only creator of life. So Islamic artists developed a special type of decoration called *arabesque*. It consists of winding stems with abstract leaves. Islamic artist stylized living things in other ways to make them look like they weren't actually alive. For example, they would often present figures incompletely, or make them appear faceless, or make them look like flattened shapes.

Another important development in Islamic art was the use of Arabic script, which lends itself to beautiful writing called *calligraphy*. Islamic artists often used calligraphy to decorate art objects and walls of religious buildings with writings from the *Qur'ān,* the holy book of Islam.

Islamic artists found their greatest expression in architecture, especially *mosques*. Mosques are Islamic houses of worship. The city of Isfahan features one of Iran's most magnificent mosques—*Masjid-e-Imam* (Imam Mosque). The Mosque of Shaykh Lutfullah, with glazed tilework adorning its elaborate entrance and dome, is also in Isfahan.

Iraq is an Arab country situated at the head of the Persian Gulf. It is almost entirely landlocked. Iraq's only outlet to the Persian Gulf is through the Shatt al Arab, a waterway formed by the meeting of the Tigris and Euphrates rivers. The country's most important port, Basra, stands on the banks of the Shatt al Arab, about 55 miles (90 kilometers) from the Persian Gulf.

Iraq's history began more than 5,000 years ago in the Tigris-Euphrates Valley. Here, the Sumerians developed the world's first civilization. Sumer was an ancient region in southern Mesopotamia, which is now southeastern Iraq. The Sumerian civilization began about 3500 B.C. The Sumerians built magnificent palaces and temples. They also had knowledge of mathematics, astronomy, and medicine. The Sumerians invented the world's first writing system. The system began chiefly as a set of word pictures and developed into a script called *cuneiform*.

Built on the ruins of the world's first civilization, Iraq has developed into one of the leading oil-producing nations in the world. However, a series of violent conflicts has caused great troubles for the nation since the late 1900's.

From 1980 to 2003, Iraq—under the leadership of President Saddam Hussein—fought three wars that have had devastating effects on the country. In the first war, Iraq fought Iran from 1980 to 1988. The second war followed Iraq's invasion and occupation of Kuwait in 1990. In response to the invasion, 39 nations sent forces to the region and defeated Iraq in the Persian Gulf War of 1991. In 2003, a military coalition led by United States forces invaded Iraq and overthrew Hussein. The war officially ended in 2011.

A new Iraqi government took office after the overthrow of Hussein, but the country was left unstable and on the brink of civil war. Iraqi and foreign militants, claiming to fight for Iraq's freedom from invaders, carried out many attacks against coalition troops and their Iraqi allies, both military and civilian.

Ancient empires

The Sumerians were the first civilization to live in Iraq. Their advanced society thrived there until about 2000 B.C. The Sumerian region centered on the great city of the ancient world, Babylon.

IRAQ

Archaeological records show the first mention of Babylon around 2200 B.C. Babylon was the capital of the kingdom of Babylonia and of two Babylonian empires. The present-day Iraqi city of Al Hillah stands on the site of Babylon.

Babylon was famed for its developments in learning, architecture, sculpture, mathematics, and geometry. Its culture strongly influenced ancient Greek philosophers. The Hanging Gardens of Babylon were one of the Seven Wonders of the Ancient World. These gardens, laid out on elevated brick terraces, were probably built by King Nebuchadnezzar II for his wife. The Babylonian Empire lasted until 539 B.C., when Persia conquered the region and added it to their empire.

When Alexander the Great defeated the Persians in 331 B.C., he began a period of Greek rule. The region became part of the Roman Empire in A.D. 115.

The Arabs spread through Mesopotamia in A.D. 637, bringing the Arabic language and the Islamic religion to the region. Baghdad was the capital of the Arab Empire, and it became a great center of learning. The town of Samarra also served briefly as the Arab capital in the 800's.

During the 1200's, Baghdad was destroyed in a Mongol invasion of the region. This area never regained its power or importance after the Mongol conquest. It became part of the Ottoman Empire when the Ottoman Turks seized Mesopotamia in 1534.

Modern Iraq

The Turks ruled the Mesopotamian region for almost 400 years. During World War I (1914-1918), troops from the United Kingdom took Mesopotamia from the Ottomans. In 1920, the League of Nations gave the British administrative rule in Mesopotamia. The British helped the region's leaders set up a government in 1921. These leaders named the country Iraq and elected King Faisal I the first monarch.

Iraq gained its independence in 1932. During the following years, a series of military revolts ended the monarchy, and the revolutionaries declared Iraq a republic in 1958.

187

IRAQ TODAY

The history of Iraq since the 1960's has been dominated by political turmoil and bloody conflicts. In 1963, the rebels who had established Iraq as a republic were overthrown by Abdul Salam Arif and Ahmad Hasan al-Bakr, who were members of the Baath Party. The party's philosophy blended economic socialism and Arab nationalism. In 1968, al-Bakr set up a Baathist-controlled military government in Iraq. In 1979, Bakr resigned from office, and Saddam Hussein succeeded him.

War against Iran

In 1980, Iraq entered into a war against Iran that proved to be very harmful to Iraq. Many Iraqi soldiers and civilians were killed. Missiles struck Baghdad and other cities. Iraq's normal trade routes were disrupted, and its ports were closed. In August 1988, both sides agreed to a cease-fire.

The Persian Gulf War of 1991

In August 1990, Iraqi forces under Hussein's leadership invaded and occupied Kuwait. Hussein hoped that, by annexing Kuwait as an Iraqi province, Iraq could absorb the country's wealth. Iraq's economy had been badly damaged in the war with Iran.

However, the United Nations (UN) Security Council called for Iraq to withdraw from Kuwait. In November 1990, the UN approved the use of force to remove Iraqi troops from Kuwait if they did not leave Kuwait by Jan. 15, 1991. Iraq refused to withdraw, and international forces led by the United States attacked military and industrial targets in Iraq and Kuwait. Later, allied forces launched a ground attack into Kuwait, quickly defeating the Iraqi troops.

Many Iraqi soldiers and civilians died as a result of the war, and bombing severely damaged Iraq's transportation systems, communication systems, utilities, and many industries. The UN imposed sanctions that banned Iraq from selling oil, its main revenue source.

In November 1994, Iraq formally recognized the independence of Kuwait. Also in 1994, fighting broke out between rival groups of Kurds in northern Iraq. In 1996, the Iraqi government sent troops to support one of the Kurdish groups. The United States opposed this action and launched missiles against southern Iraq.

FACTS

Official name:	Al Jumhuriyah al Iraqiyah (State of Iraq)
Capital:	Baghdad
Terrain:	Mostly broad plains; reedy marshes along Iranian border in south with large flooded areas; mountains along borders with Iran and Turkey
Area:	169,235 mi^2 (438,317 km^2)
Climate:	Mostly desert; mild to cool winters with dry, hot, cloudless summers; northern mountainous regions experience cold winters with occasional heavy snows that melt in early spring, sometimes causing extensive flooding in central and southern Iraq
Main rivers:	Euphrates, Tigris, Little and Great Zab
Highest elevation:	About 11,840 ft (3,609 m) in Zagros Mountains
Lowest elevation:	Persian Gulf, sea level
Form of government:	Federal republic
Head of state:	President
Head of government:	Prime minister
Administrative areas:	18 muhafazat (governorates)
Legislature:	Majlis Watani (Council of Representatives) with 275 members serving four-year terms
Court system:	Supreme Court
Armed forces:	577,000 troops
National holiday:	Republic Day - July 14 (1958)
Estimated 2010 population:	30,623,000
Population density:	181 persons per mi^2 (70 per km^2)
Population distribution:	66% urban, 34% rural
Life expectancy in years:	Male, 62; female, 66
Doctors per 1,000 people:	0.7
Birth rate per 1,000:	32
Death rate per 1,000:	7
Infant mortality:	44 deaths per 1,000 live births
Age structure:	0-14: 41%; 15-64: 56%; 65 and over: 3%
Internet users per 100 people:	0.9
Internet code:	.iq
Languages spoken:	Arabic (official), Kurdish (official in Kurdish regions), Assyrian, Armenian
Religions:	Muslim 97% (mainly Shiah), Christian 3%, others less than 1%
Currency:	New Iraqi dinar
Gross domestic product (GDP) in 2008:	$92.35 billion U.S.
Real annual growth rate (2008):	9.8%
GDP per capita (2008):	$2,983 U.S.
Goods exported:	Mostly: crude oil Also: aluminum, copper, food, machinery
Goods imported:	Electronics, food, machinery, motor vehicles
Trading partners:	Italy, South Korea, Syria, Turkey, United States

The Iraq War

In 1996, the UN partially lifted the embargo on Iraq, allowing it to export oil on a limited basis. In 1998, Iraq began to refuse the entry of UN weapons inspectors into the country. In December of that year, the United States and the United Kingdom launched air strikes against Iraq. United States and British officials said the attacks were to limit Iraq's ability to make weapons of mass destruction (biological, chemical, and nuclear weapons).

In 2002, under threat of military attack by the United States, Iraq allowed UN inspectors to return. During the inspections, the United States continued to accuse Iraq of violating UN disarmament terms.

In March 2003, U.S.-led forces launched an air attack against Baghdad. The United States declared an end to major combat on May 1, 2003. Fighting continued, however. Saddam Hussein was captured by U.S. troops in December 2003. After a trial, Iraqi authorities executed Hussein in 2006.

In 2004, an interim government took power. Later that year, a team of U.S. and British experts stated that no weapons of mass destruction had been found in Iraq. In 2005, voters elected a transitional National Assembly, which oversaw the preparation of a constitution. Iraqis approved the constitution in a nationwide referendum and then elected a new government. Nouri Kamel al-Maliki became prime minister in 2006. He continued as prime minister after elections in 2010.

Fighting continued in Iraq after the capture of Hussein. Clashes between U.S.-led troops and Sunni and Shi`ite militants took place throughout Iraq. In 2007, the United States sent 30,000 more troops to help the Iraqi government secure the country. Combat operations ended on Aug. 31, 2010, and by the end of 2011, all U.S. troops had withdrawn from Iraq.

In 2011, antigovernment protests erupted in several Iraqi cities. The protesters called for improved government services and an end to corruption. Several people were killed in clashes with security forces.

American soldiers prepare to enter a building in Baqubah, northeast of Baghdad. The troops were trying to end ongoing violence in the area. U.S.-led forces invaded Iraq in 2003.

ENVIRONMENT

Iraq, which covers an area slightly larger than the state of California, can be divided into four land regions. They are (1) the upper plain, (2) the lower plain, (3) the mountains, and (4) the desert.

The upper plain

The upper plain is made up of the region that lies between the Tigris and Euphrates rivers north of the city of Samarra. The upper plain is mostly dry, rolling grassland. The highest hills in the region rise about 1,000 feet (300 meters) above sea level. The Tigris and Euphrates flow swiftly through the upper plain, carrying large amounts of sediment.

Mosul is the largest city in the upper plain, with about 1.5 million inhabitants. It is located in the northern sector of the region, about 60 miles (100 kilometers) from the Syrian and Turkish borders. Since 1952, Mosul has been the center of an important oil field.

The lower plain

The lower plain extends from Samarra to the Persian Gulf. It includes the area between the Tigris and the Euphrates. Although the lower plain receives little rainfall, the farmlands are green with vegetation. An elaborate system of irrigation canals, first dug centuries ago, brings water to the crops.

Farmers also rely on regular flooding of the region to water their crops. Floods occur in April and May each year, when the Tigris and Euphrates overflow their banks.

Old-fashioned water wheels still stand on the river banks. Here, palm, orange, and lemon groves have provided abundant crops of fruit for centuries.

Modern methods of irrigation allow Iraqi farmers to extend their fields into the drier areas. As a result, they now have enough water to grow wheat, cotton, rice, tomatoes, melons, and grapes. Huge groves of date palms lie south of Baghdad.

Along the Shatt al Arab waterway, date palms grow in large groves. The date palm is one of the oldest crop plants. According to Muslim legend, it represents the tree of life. A date palm can grow as tall as 100 feet (30 meters).

The northern areas of Iraq are also known as Kurdistan. They make up the foothills and valleys of the Zagros Mountains. These forested valleys and cool uplands are quite different from the sands and stony plains of the adjoining desert.

This region is renowned for its tall palm trees crowned in the autumn with ripening fruit. Dates rank among Iraq's valuable exports.

In the south, where the Tigris and Euphrates meet to form the Shatt al Arab, the land becomes a vast swamp with two marshy lakes, the Hawr al Hammar and the Hawr as Saniyah. Several natural channels link the isolated settlements of Iraq's *madan* (Marsh Arabs). The madan live in reed houses built on artificially made islands of mud and reeds. They paddle from one island to another in canoes.

In 1993, Iraq's government began to drain these marshlands to punish the Marsh Arabs who supported an uprising against Saddam Hussein following the Persian Gulf War of 1991. Many residents had to leave the area to find food. In 2003, efforts to restore the marshlands were begun, and former residents began to return to the area.

The mountains

The mountains of northeastern Iraq form part of the range called Zagros in Iran and Taurus in Turkey. Some Iraqi peaks in this range rise to more than 11,000 feet (3,350 meters) above sea level.

The region is generally cold and wet. In winter, many of the mountain valleys are cut off from each other by heavy snow. In the milder months, spectacular waterfalls, forests of walnut trees, and refreshing cold springs attract thousands of visitors.

The foothills and valley region are also called *Kurdistan* because the area is home to the Kurds. Many Kurds farm the land and herd sheep and goats for a living.

The desert

The desert covers southern and western Iraq. It also extends into Jordan, Kuwait, Saudi Arabia, and Syria. Most of this region of limestone hills and sand dunes is part of the Syrian Desert. Much of the land is almost completely uninhabited, but some villages and small towns can be seen near the old caravan trails. In recent years, highways have replaced the camel trails across the desert.

Tribes of Bedouin herdsmen still drive their sheep through the desert. Some tribes have abandoned the nomadic way of life to settle in agricultural communities.

Naour (water wheels) on the Euphrates provide necessary irrigation for Iraqi agriculture. These water systems date from the earliest civilizations. Modern concrete canals now supplement the water wheels, enabling farmers to extend their crops into drier areas.

PEOPLE

A variety of ethnic and religious groups live in Iraq. About 75 percent of the population are Arabs. Kurds make up another 20 percent. The remainder includes Armenians, Assyrians, Turkomans, and Yazidis. About 40 percent of Iraq's population is under 15 years of age. Although Arabic is the official language, many people speak Kurdish or Turkish. Quite a few are fluent in English, the international language of oil and commerce.

Religious life

More than 95 percent of Iraqis are Muslims. The majority belong to the Shiah branch of Islam. Most of the Shiites live in the central and southeastern regions of the country. The members of the other main Muslim sect, the Sunnis, live mainly in the north. Most Kurds are Sunni Muslims. Christians and other groups make up less than 5 percent of Iraq's population.

One of the holiest Shiite shrines stands in the city of Karbala, some 50 miles (100 kilometers) south of Baghdad. For centuries, pilgrims from all over the world have come there to visit the shrine of Imam Husayn, a Shiite saint who lived in the 600's. Women must wear a veil when crossing the square in front of the mosque and when walking through the town bazaar.

For decades, the Shiites have felt excluded from positions of national power and prestige. Their angry feelings began in the 1930's. When the Baath Party ruled Iraq from 1968 to 2003, most of its high-ranking members were Arab Sunni Muslims. Many Shiites resented the Sunni monopoly on government power. During the Iran-Iraq war of the 1980's, Iranians hoped that a Shiite rebellion within Iraq would weaken their enemy. They were surprised to find that the Iraqi Shiites supported their government and the war effort.

In 2005, Iraq set up a government that allows Shiites, Sunnis, and Kurds to share political power.

The new and the old

Religious differences are not the only conflicts that split Iraq today. In parts of Baghdad, modern ways often come up against age-old traditions.

Baghdad's *suq*, or bazaar, is one of the few places in the city that has escaped modernization. For centuries, rich and poor alike have shopped the covered market for everything from spices to the wares of the goldsmiths and coppersmiths.

Shiite pilgrims march through a cemetery in Najaf on their way to the city of Karbala. Shiites make an annual pilgrimage to Karbala to mark the end of 40 days of mourning that follow the anniversary of the death in the A.D. 600's of Imam Husayn, a grandson of the Prophet Muhammad.

A street vendor outside one of Baghdad's mosques wears the traditional headdress and baggy trousers of the Kurdish people. Iraq has the largest population of Kurds in the Middle East. They account for about 20 percent of Iraq's total population. Most Kurds live in the foothills and valleys of the mountains in northeastern Iraq. This region is called Kurdistan.

In contrast to the bazaar, modern luxury hotels line the banks of the Tigris River in Baghdad. These hotels draw the country's elite, many of whom are attracted to the Western lifestyle.

The differences between the new and the old can be seen in the position of women in Iraq today. It is not unusual to see a modestly veiled Iraqi mother with a daughter who wears Western fashions and cosmetics. Since the late 1900's, women have made up a growing percentage of Iraq's work force.

The discontent of the Kurds is another source of tension in Iraq. Many Kurds would like to have their own separate Kurdish state. They regularly stage demonstrations to express their wish for political independence. In 1970, the Baath Party offered the Kurds radio and television programs in their own language, as well as Kurdish instruction in many schools. But for many Kurds, that was not enough. They took sides with Iran against the Iraqi army in the Iran-Iraq war. After the war ended, the Kurds suffered harsh treatment at the hands of the Iraqi army.

The Persian Gulf War of 1991 triggered another Kurdish uprising in northern Iraq, but the Iraqi army swiftly put down the rebellion. More than a million Kurds then fled to the mountains, where thousands of them died of disease, exposure, or hunger. Conditions for the Kurds improved after the Iraq War (2003-2011). In 2005, Jalal Talabani became Iraq's first Kurdish president. He was reelected in 2010.

A small group of people who speak Turkish live in the northern part of Iraq. The Turkish-Iraqi border region is also home to the descendants of the ancient civilization of Assyrians and Yazidis, who are of Kurdish origin.

An Iraqi vendor sells pickles to a customer at the Al-Shorja market in central Baghdad. The market, dating back 200 years, is the oldest in the city. This market is one of many scattered throughout the city that serve local residents.

ECONOMY

Iraq's economy is built mainly around oil. The export of oil has played an important economic role in Iraq since the 1950's. Iraq's government used oil profits to strengthen other areas of the economy, such as manufacturing and agriculture, in the 1970's. But wars in the 1980's, 1990's, and 2000's have caused great damage to all sectors of Iraq's economy.

Oil and industry

Iraq is a member of the Organization of the Petroleum Exporting Countries (OPEC). Before OPEC was founded in 1960, the petroleum industry in Iraq and other Middle Eastern countries was controlled by oil companies in the United States and Europe.

Countries like Iraq were paid royalties on the oil they produced. These royalties were based on the *posted price* set by the oil companies for crude oil. OPEC gave oil-producing nations more control over oil pricing and production.

In the 1970's, worldwide demand for oil was greater than the supply that could be produced by non-OPEC countries. OPEC nations, like Iraq, then dramatically raised their prices for crude oil.

This price increase created an economic boom in OPEC nations. In Iraq, in addition to modernizing agriculture, the government used oil profits to finance a rapid expansion of the country's steel, plastics, cement, and fertilizer industries. By the end of the 1970's, oil revenues had also established Iraq's textile and electronics industries and supported many public works projects, including hospitals, schools, roads, and airports.

The outbreak of the war against Iran in 1980 ended Iraq's oil boom, along with its dream of quick and easy profits. Just weeks after the war began, Iranian commandos destroyed Iraq's oil-loading facilities along the shores of the Persian Gulf. The nation's oil exports slowed to a trickle. Iraq's industries also suffered, and plans for building new cities and factories had to be postponed. Iraq's dependence on profits from crude oil led to severe financial losses during the war, as normal trade routes were interrupted.

Iraqis walk past the damaged al-Askari shrine in Samarra. The shrine was bombed in 2006 during ongoing religious-inspired violence within the country in the early 2000's.

Iraq's invasion of Kuwait in August 1990 caused another severe blow to the country's economy. Four days after the invasion, the United Nations (UN) Security Council imposed an embargo that prohibited all trade with Iraq except for medical supplies. Since nearly all of Iraq's major trading partners supported the embargo, Iraq's foreign trade all but ended.

After the Persian Gulf War of 1991, the UN continued the embargo to pressure Iraq into carrying out the terms of the formal cease-fire agreement. Under these terms, Iraq was to destroy all of its biological and chemical weapons and repay Kuwait for war damages. The embargo was eased in 1996 to allow Iraq to export oil in exchange for food and other nonmilitary goods.

Iraq's oil-producing capability was damaged further by the Iraq War, which began in 2003 and ended in 2011. The UN trade embargo was lifted in mid-2003, allowing oil exports to resume fully. Today, in addition to oil, natural gas and salt rank among Iraq's leading mined products.

Agriculture and manufacturing

About one-third of Iraq's people make their living from farming and herding. Iraqi farmers grow barley, dates, grapes, melons, olives, oranges, rice, tomatoes, and wheat. Herders raise camels, goats, horses, and water buffalo.

Iraq's manufacturing industries employ a relatively small percentage of the population. Oil refining and petrochemical production make up an important industry, despite wartime damage to refineries. Several of Iraq's chemical and oil plants are near the cities of Baiji, Basra, and Kirkuk. Other factories in Iraq process farm products or make such goods as cement, fertilizers, and textiles.

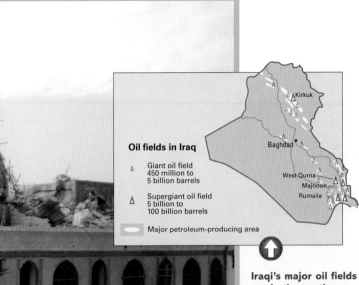

Oil fields in Iraq

Kirkuk
Baghdad
West Qurna
Majnoon
Rumaila

△ Giant oil field
450 million to
5 billion barrels

△ Supergiant oil field
5 billion to
100 billion barrels

Major petroleum-producing area

Iraqi's major oil fields are in the southern part of the country near the Kuwait border and near Kirkuk in the north. Oil is Iraq's chief mineral resource. Oil refining and petrochemical production make up an important industry.

Unwanted gas burns off in an oil field near Kirkuk. Several of Iraq's oil and chemical plants are located near Kirkuk as well as near the cities of Baiji and Basra.

Farmers grow cucumbers under plastic covers near Samarra, along the Tigris River north of Baghdad.

MESOPOTAMIA

The valleys of the Tigris and Euphrates rivers in present-day southern Iraq were home to the world's earliest civilizations. They were the center of an ancient region called Mesopotamia—a Greek word meaning *between the rivers*. Mesopotamia included the area that is now eastern Syria, southeastern Turkey, and most of Iraq.

In spring, water from melting snow on Iran's high plateau causes the Tigris and Euphrates rivers to overflow their banks. Through the centuries, the regular flooding of the valley between the rivers created vast marshlands. Sometime before 3500 B.C., a group of people settled in these marshlands and developed the first civilization. This area came to be called Sumer.

The marshes where the Sumerians first settled are natural swamplands. The main vegetation is a reed called *qasab*.

The Sumerians built irrigation ditches and canals that allowed water to be diverted to areas away from the rivers. Their irrigation system gave the Sumerians rich harvests of barley, wheat, dates, and vegetables. It also saved labor, which meant that people had time and energy for other activities.

The Sumerians' most important achievement was inventing the world's first writing system. From this first system of word pictures, the Sumerians developed a script called *cuneiform*. The Sumerians made cuneiform characters by pressing a tool with a wedge-shaped tip into wet clay tablets and then drying the tablets in the sun. Hundreds of thousands of these tablets have been found by archaeologists. They indicate that these ancient people had knowledge of mathematics, astronomy, and medicine.

By about 3500 B.C., the Sumerians had established several cities. These cities eventually grew into independent city-states. The more powerful city-states conquered their neighbors and set up independent kingdoms.

A reconstruction of a Sumerian temple in the ancient city of Ur shows how these ancient people built temples called ziggurats in stepped formation.

Marsh Arabs have inhabited the the swamplands of Mesopotamia for about 5,000 years. Saddam Hussein ordered dams built and diverted waterways to drain the marshes in order to destroy the Marsh Arabs. But some dams have been reopened, allowing the marshes slowly to refill. As a result, Marsh Arabs began returning to their traditional lifestyle.

A Marsh Arab paddles his canoe down a waterway near Qurna. Marsh Arabs use the waterways to transport goods and people.

Wars between the city-states brought the downfall of Sumer during the 2300's B.C. The Akkadians, a Semitic people from the west, conquered Sumer and, along with other groups, ruled Mesopotamia between 2300 and 539 B.C. In 539 B.C., Mesopotamia became part of the Persian Empire. Alexander the Great conquered the Persians in 331 B.C.

Around A.D. 750, Islam's Abbasid dynasty moved the Islamic capital from Damascus to Baghdad, on the Tigris River. Then, in 1258, the Mongols crushed the Abbasids and almost destroyed Baghdad. During this period, the marshes of Lower Mesopotamia became a stronghold of the Zanj, a people once used as slaves to drain the marshes around Basra. In 1534, the Ottoman Turks seized the region.

Mesopotamia remained part of the Ottoman Empire until World War I (1914-1918), when the British took control. In 1921, most of Mesopotamia became part of Iraq.

Marsh Arabs still inhabit the region once known as Lower Mesopotamia. They live in reed houses and have their own unique culture.

The world's earliest civilization developed in Mesopotamia. Most of this region now lies in Iraq, with some territory lying in Syria and Turkey.

THE KURDS

The Kurds, a tribal society of farmers and herders, live in the mountainous regions of southwest Asia. Their homeland, called Kurdistan, is officially only a small province in Iran. Yet the name *Kurdistan* is generally given to the area stretching from the southern border of Azerbaijan through the Pontic and Taurus ranges in eastern Turkey, across northeast Syria and Iraq to the Zagros Mountains in northwestern Iran.

The number of Kurds living within this large area is generally estimated at about 25 million. In Iraq, Kurds make up about 20 percent of the total population. In Turkey, they are the largest minority. The major Kurdish cities are Mahabad, Sanandaj, and Kermanshah in Iran; Irbil, Kirkuk, and As Sulaymaniyah in Iraq; and Diyarbakir and Van in Turkey.

The majority of Kurds are Sunni Muslims. They speak Kurdish, an Indo-European language closely related to Persian.

History of the Kurds

The origins of the Kurds are uncertain. Historians believe that the mountains above Mesopotamia were settled around the 600's B.C. by a tribal people who fought—and sometimes defeated—the people of the plains. These tribal people were probably the early ancestors of the Kurds.

In the 1100's, the Seljuk Sultan Sandjar created a large province for these people and called it Kurdistan. Until the 1500's, the Kurdish tribes were under the control of the various powers who had conquered that part of southwest Asia. In 1514, the Medes defeated the Assyrian Empire, and Kurdistan became a separate country between the Turkish and Persian empires.

But to this day, the Kurds have never had their own government. And because they want to be politically independent, they have often been in conflict with the governments under which they live.

Kurdistan is the homeland of a tribal society called the Kurds. They live in rural communities in a mountainous region where they farm and herd sheep and goats.

Iraqi Kurds, like this man carrying loaves of flat bread, make up about 20 percent of the country's population. Most Kurds are Sunni Muslims.

THE KURDISH STRUGGLE

Since the beginning of the 1900's, many Kurds have left their rural life in the mountains for jobs in the cities. Kurds have settled in the cities of Kermanshah in Iran, Kirkuk and Irbil in Iraq, and Diyarbakir in Turkey. There, they have continued the struggle for Kurdish independence. The struggle has its roots deep in history. Since the 1920's, there have been several major conflicts between the Kurds and the governments of the countries in which they live. After World War I (1914-1918), the Turkish-Allied Treaty of Sèvres (1920) promised an independent Kurdistan. However, Kemal Atatürk, leader of the Turkish Nationalists, rejected the Treaty of Sèvres. He signed a new peace treaty with the Allies, the Treaty of Lausanne, which made no mention of the Kurds.

The Kurds have since staged a number of revolts. Under the leadership of Mustafa Barzani, Iraqi Kurds rose up in force in 1961. A truce was reached in 1970, and Iraq promised the Kurds limited self-government. But the agreement ended in a dispute over oil rights,

since the Kurdish homeland has some of Iraq's richest oil reserves. After the Iran-Iraq war ended in 1988, the Kurds suffered harsh treatment from Iraqi troops for their support of Iran during the conflict. After the Persian Gulf War of 1991, the Kurds again rebelled, but the revolt was crushed by Iraqi President Saddam Hussein. Some 1 million Kurds fled to the mountains. There, thousands died of exposure, hunger, and disease. Throughout their conflict with Iraq, Kurdish refugees have fled to Turkey for safety and protection, where they often live in primitive camps.

The Iraq War (2003-2011) brought new challenges. In April 2005, an interim parliament elected Kurdish leader Jalal Talabani president of Iraq. He became the first Kurdish president in the history of Iraq and the first Kurdish head of an Arab state. Talabani was elected to a full four-year term as president in 2006 and reelected in 2010.

Kurdish uprisings in Iran followed the revolution of 1979, and fighting also broke out in Turkey. In March 1995, Turkey invaded northern Iraq with the aim of destroying camps used by Kurdish guerrillas who had committed acts of violence within Turkey.

Kurdistan is a vast area of mountains and plateaus. It extends from Turkey across Syria and Iraq through northwestern Iran to the Azerbaijan border.

Way of life

Most Kurds are farmers and herders. The farmers grow cotton, tobacco, and sugar beets. The herders raise sheep and goats.

Kurdish farming communities are led by an *agha*, who owns the village and the farmland. The farmers grow crops as their ancestors did, using wooden plows drawn by oxen. They live in simple houses made of mud bricks. The interiors are brightened by the handwoven rugs for which the Kurds are famous. Many Kurds still wear their colorful, embroidered traditional dress.

Kurdish women enjoy much more freedom than women in most Islamic societies. They do not wear the traditional veil, and they often hold important positions within their communities.

IRELAND

Ireland, a small, independent country in northwestern Europe, covers about five-sixths of the island of Ireland. Dublin is its capital and largest city. The northeastern corner of the island, called Northern Ireland, is part of the United Kingdom of Great Britain and Northern Ireland.

Ireland is sometimes called the *Emerald Isle* because of its beautiful green countryside. Gently rolling farmland covers most of central Ireland, and mountains rise on its coasts. The picturesque ports of Cork and Waterford lie on natural harbors on the south coast, and Dublin Bay, on the east, serves as a harbor for Dublin, Ireland's major port. Many inlets and bays cut deeply into the more rugged west coast. Among them are the beautiful Galway Bay and the mouth of the River Shannon. Because of these inlets and bays, no point in the country is more than 70 miles (110 kilometers) from the sea.

The Irish are known for their friendliness and hospitality, as well as their skill in storytelling and writing. Ireland has produced a remarkable number of famous writers, including William Butler Yeats, Oscar Wilde, George Bernard Shaw, James Joyce, and Samuel Beckett. Irish folk music, known around the world, ranges from lively jigs to age-old songs of love and sorrow and the emigrant's longing for home.

Saint Patrick, the *patron*—or guardian—saint of Ireland, brought Christianity to the country in the A.D. 400's. Today, more than 85 percent of the Irish people are Roman Catholics. For hundreds of years, the Irish Catholics were persecuted by the Protestant British. In 1919, Catholic Ireland declared its independence from the United Kingdom. In 1921, a treaty was signed that created an independent Ireland with its present borders. Predominantly Protestant Northern Ireland remained part of the United Kingdom. Many people in Ireland and many Catholics in Northern Ireland want their states to reunite. However, the majority of the people of Northern Ireland want the lands to remain separate.

Emigration has been a major problem in Ireland. The potato famines of the 1840's forced hundreds of thousands of people to leave the country. In the early 1900's, thousands of people emigrated because of the limited job opportunities in Ireland. The development of new industries helped check emigration in the mid-1900's, and by the 1990's, Ireland had one of the strongest economies in Europe. Many people who had left returned. In the early 2000's, a world-wide economic crisis caused Ireland's economy to fall into recession, and emigration rose to its highest level in 20 years.

IRELAND TODAY

Ireland is a republic with a president, a prime minister known as the *taoiseach*, and a parliament. The president, the official head of state, is elected by the people to a seven-year term and may not serve for more than two terms. The powers of the president are limited to appointing the taoiseach, approving laws, and calling Parliament into session.

The taoiseach, the head of government, is usually the leader of the majority party in the legislature's lower house. The taoiseach selects other members of Parliament to serve in the Cabinet.

Parliament, called the *Oireachtas* in the Irish language, consists of the president, the legislature's lower house *(Dáil Éireann),* and the Senate *(Seanad Éireann).* The Dáil, which has 166 members elected by the people, makes Ireland's laws. The Senate serves mainly in an advisory capacity. Some senators are elected and some are appointed.

Ireland's main political parties include *Fianna Fáil* (Soldiers of Destiny), *Fine Gael* (Gaelic People), the Labour Party, the Democratic Left, and the Progressive Democratic Party. All Irish citizens who are at least 18 years old can vote at general elections. British citizens living in Ireland also are able to vote in Dáil and local elections, and European Union citizens can now vote in local government and European parliamentary elections.

On the local level, Ireland is divided into 26 counties and 5 county boroughs. Each county and county borough is governed by an elected council and a county manager appointed by the national government.

About 60 percent of Ireland's people live in cities and large towns, although only Dublin and Cork have more than 100,000 inhabitants. The rest of the Irish people live in small rural towns and villages or in the countryside.

Ireland has two official languages—English and Irish, a Celtic language. All the people speak English. Although about 40 percent consider themselves able to converse in Irish, only a small number use it as their

FACTS

Official name:	Ireland
Capital:	Dublin
Terrain:	Mostly level to rolling interior plain surrounded by rugged hills and low mountains; sea cliffs on west coast
Area:	27,133 mi² (70,273 km²)
Climate:	Temperate maritime; modified by North Atlantic Current; mild winters, cool summers; consistently humid; overcast about half the time
Main rivers:	Shannon, Liffey, Barrow, Boyne, Moy, Nore, Suir
Highest elevation:	Carrauntoohil, 3,414 ft (1,041 m)
Lowest elevation:	Atlantic Ocean, sea level
Form of government:	Republic
Head of state:	President
Head of government:	Prime minister
Administrative areas:	26 counties
Legislature:	Oireachtas (Parliament) consisting of the Seanad Éireann (Senate) with 60 members serving five-year terms and the Dáil Éireann (House of Representatives) with 166 members serving five-year terms
Court system:	Supreme Court
Armed forces:	10,500 troops
National holiday:	St. Patrick's Day - March 17
Estimated 2010 population:	4,458,000
Population density:	164 persons per mi² (63 per km²)
Population distribution:	61% urban, 39% rural
Life expectancy in years:	Male, 76; female, 82
Doctors per 1,000 people:	2.9
Birth rate per 1,000:	16
Death rate per 1,000:	6
Infant mortality:	4 deaths per 1,000 live births
Age structure:	0-14: 21%; 15-64: 68%; 65 and over: 11%
Internet users per 100 people:	64
Internet code:	.ie
Languages spoken:	English (official), Irish (official)
Religions:	Roman Catholic 87.4%, Church of Ireland 2.9%, other Christian 1.9%, other 7.8%
Currency:	Euro
Gross domestic product (GDP) in 2008:	$280.04 billion U.S.
Real annual growth rate (2008):	-1.7%
GDP per capita (2008):	$65,597 U.S.
Goods exported:	Chemicals, computers and other office machinery, meat and dairy products, pharmaceuticals
Goods imported:	Chemicals, computer and office machine parts, crude oil and petroleum products, food, motor vehicles
Trading partners:	Belgium, France, Germany, United Kingdom, United States

The following is a list of place names and features shown on the map of Ireland:

Norway, Inishtrahull, Malin Head, SCOTLAND, Fanad Head, Atlantic Ocean, Denmark, Inishowen Peninsula, Inishowen Head, Buncrana, North Channel, North Sea, United Kingdom, Netherlands, IRELAND, EUROPE, Belgium, France

Sheep Haven, Tory I., Bloody Foreland, GLENVEAGH N.P., Mt. Errigal 2,466 ft (752 m), Aran I., Donegal, Ballybofey, Letterkenny, Lifford, NORTHERN IRELAND, Lough Swilly, Lough Foyle

Gweebarra Bay, Dawros Head, Rossan Point, Blue Stack Mtn. 2,219 ft (676 m), Carrigan Head, Donegal Bay

Broad Haven, Erris Head, Downpatrick Head, Killala Bay, Sligo Bay, Inishmurray, L. Melvin, Sligo, Cuilcagh Mtn., Monaghan

Mullet Peninsula, Carrowmore L., Ballina, Sligo, Leitrim, L. Gill, Iron Mts., Monaghan, Muckno L., Carlingford Lough, Dundalk, Louth

Inishkea, Blacksod Bay, Mountains of Mayo, Ox Mts., L. Conn, Cavan, Cavan, Ardee, Dunany Point

Achill Head, Achill I., Castlebar, Mayo, L. Mask, Moy, L. Bofin, Carrick-on-Shannon, Clogher Head, Drogheda, Laytown

Clare I., Clew Bay, Westport, Roscommon, Longford, Meath, Navan, Tara Hill 509 ft (155 m), Skerries, Rush, Irish Sea

Inishturk, Mweelrea 2,688 ft (819 m), L. Carra, Roscommon, Longford, Westmeath, Dunshaughlin, Maynooth, Lambay I., Malahide, Irelands Eye

Inishbofin, Inishark, CONNEMARA N.P., Mountains of Connemara, L. Corrib, Galway, Mullingar, Edenderry, Swords, Dublin, Dublin Bay

Slyne Head, Bertraghboy Bay, Iar Connaught, Galway, Athlone, Ballinasloe, Tullamore, Kildare, Droichead Nua, Naas, Leixlip, Dublin, Dún Laoghaire, Bray

North Atlantic Ocean, Gorumna I., Inishmore, Aran Islands, Inishmaan, Inisheer, CLIFFS OF MOHER, Hags Head, Loughrea, Offaly, Portarlington, Ardern +1,733 ft (528 m), Port Laoise, Kildare, Athy, Greystones, WICKLOW MOUNTAINS N.P., Wicklow, Wicklow Head

Liscannor Bay, Mal Bay, Burren, BURREN N.P., Slieve Aughty Mts., Clare, Slieve Bernagh, Roscrea, Laois, Lugnaquillia Mtn. 3,039 ft (926 m), Arklow

Donegal Point, Slievecallan 1,282 ft (391 m), Ennis, Nenagh, Keeper Hill +2,279 ft (694 m), Thurles, Slieveardagh Hills, Kilkenny, Carlow, Carlow, Kilmichael Point

Loop Head, Mouth of the Shannon, Shannon, Limerick, Tipperary, Mt. Leinster 2,610 ft (796 m), Gorey, Cahore Point

Kerry Head, Listowel, Limerick, Newcastle West, Galty Mts., Clonmel, Kilkenny, Enniscorthy, New Ross, Wexford, Wexford Harbour

Ballyheige Bay, Stack's Mts., Mullaghareirk Mts., Carrick-on-Suir, Wexford, Greenore Point

Brandon Mountain 3,127 ft (953 m), Tralee, Mitchelstown, Knockmealdown Mountains, Comeragh Mts., Waterford, Tramore, Hook Head, Carnsore Point, St Georges Channel

Sybil Point, Slieve Mish Mts., Kerry, Blackwater, Waterford, Saltee Is.

Dingle, Dingle Bay, KILLARNEY N.P., Killarney, Nagles Mts., Dungarvan, Waterford Harbour

Great Blasket I., Carrauntoohil 3,414 ft (1,041 m), Macgillicuddy's Reeks, Boggeragh Mts., Cork, Cork, Youghal, Mine Head

Doulus Head, Valentia I., Youghal Bay, Knockadoon Head

The Skelligs, Bolus Head, Caha Mts., Passage West, Cobh, Cork Harbour

Kenmare River, Bandon, Bantry, Clonakilty, Old Head of Kinsale

Dursey I., Bantry Bay, Mt. Gabriel 1,339 ft (408 m), Clonakilty Bay

Mizen Head, Roaringwater Bay, Clear I., Mizen Head

0   25 Miles
0   25 Kilometers

IRELAND

Sligo, Dundalk, Galway, Dublin, Limerick, Tralee, Waterford, Cork

☐ Irish-language areas (Gaeltacht)

Although English is used everywhere in Ireland, Irish is the everyday language in several areas—collectively known as the Gaeltacht.

Ireland occupies five-sixths of the island of Ireland. The remainder of the island consists of Northern Ireland, which is part of the United Kingdom.

The library of Trinity College, Dublin, contains the illuminated *Book of Kells* (c. A.D. 800) made by Irish monks.

everyday language. Since Ireland gained its independence in the early 1900's, many Irish people have sought to bring the Irish language into wider use. Today, the Irish government requires schools to teach Irish as well as English and uses both English and Irish for official government business.

The Roman Catholic Church, which has long played a major role in Irish life, influences Irish law and operates many of the country's primary and secondary schools. The local church serves as a meeting place where many people take part in social events.

For hundreds of years, nearly all the Irish people made their living from farming, and many still do. However, agriculture has declined in importance since the 1920's. Today, Ireland's economy depends heavily on service industries and manufacturing.

Ireland's most famous products include *stout*, a type of beer; delicate cut glass; high-quality linen; and fine woolen clothing. The potato, which has long been important to Ireland's economy, is one of the country's major agricultural products.

HISTORY

The first people to live in Ireland probably came from the European mainland about 6000 B.C. About 400 B.C., invading Celtic tribes gained control of the island and divided it into small kingdoms.

In A.D. 432, Saint Patrick converted Ireland to Christianity. Saint Patrick also introduced the Roman alphabet and Latin literature into Ireland, and monasteries flourished there after his death. His feast day, March 17, is a national holiday in Ireland.

Around 795, the Vikings began a series of raids on Ireland, robbing and destroying the monasteries. The Irish could do little to defend themselves against the well-armed invaders until 1014, when the Irish king, Brian Boru, defeated the Vikings at Clontarf (now part of Dublin). Brian's army was victorious, but the king himself was killed.

Norman nobles began seizing Irish lands in the late 1100's. By the 1300's, Normans controlled nearly all of Ireland.

English conquest of Ireland

In 1541, Henry VIII of England, in an attempt to gain control of Ireland, forced the Irish Parliament to declare him king of Ireland. He also tried to force Protestantism on the Irish. His daughter gave counties in Ireland to English settlers, and Elizabeth I outlawed Catholic religious services. This religious persecution continued through the 1600's.

James II, a Catholic king, attempted to restore the rights of the Irish Catholics, but he was forced from the English throne in 1688. When he tried to regain power with an army raised in Ireland, Irish Protestants helped the English defeat him in 1690.

In 1801, the Act of Union made Ireland part of the United Kingdom of Great Britain and Ireland. The Irish Parliament was dissolved, but the Irish eventually won the right to send representatives to the British Parliament.

Additional misfortunes afflicted the Irish people in 1845, when a plant disease wiped out the potato crop. From 1845 to 1848, about 1 million people died of starvation or disease, and hundreds of thousands of others left the country.

Irish fishermen from the Aran Islands carry a curragh, a boat whose design dates back to prehistoric times. Early Irish settlers who moved to the island around 6000 B.C. lived by hunting and fishing.

Riflemen of the Irish Citizen Army shoot from the roof of Liberty Hall, Dublin, at the time of the Easter Rebellion of 1916. Although British troops crushed the rebellion, it inspired the Irish people to win a "Free State" five years later.

TIMELINE

c. 6000 B.C.	First people settle Ireland from Europe.
c. 400 B.C.	Celtic tribes invade Ireland.
A.D. 432	Saint Patrick converts the Irish to Christianity.
c. 795	Vikings begin a series of raids on Ireland.
1014	Vikings are defeated by Brian Boru.
1171	Normans recognize Henry II of England as lord of Ireland.
1541	Henry VIII of England is declared king of Ireland.
1690	James II and Irish forces are defeated at the Battle of the Boyne.
1801	Ireland becomes part of the United Kingdom of Great Britain and Ireland.
1845-1848	A potato famine kills about 1 million people.
1916	The Easter Rebellion against British rule breaks out in Dublin but is put down within a week.
1919	Irish republicans establish a parliament in Dublin and declare Ireland an independent republic.
1920	Northern counties of Ireland become the state of Northern Ireland.
1921	Southern Ireland becomes a dominion of the United Kingdom called the Irish Free State.
1949	The Irish Free State cuts ties with the United Kingdom and becomes a republic, Ireland.
Late 1960's	Violence again breaks out between Protestants and Catholics in Northern Ireland.
1973	Ireland joins the European Community, now known as the European Union (EU).
1985	Ireland is given an advisory role in Northern Ireland's government.
1990	Mary Robinson becomes Ireland's first female president.
1999	Ireland begins taking part in new governing bodies created by the Northern Ireland peace agreement.
2010	The collapse of several Irish banks pushes the government into near-default, plunging Ireland into severe recession.

Jonathan Swift,
satirist
(1667-1745)

Daniel O'Connell,
political reformer
(1775-1847)

William Butler Yeats,
poet
(1865-1939)

After Saint Patrick
preached Christianity
to the Celts in Ireland,
the Roman Catholic
monasteries in Ireland
were a stronghold of
Western civilization
and art during
Europe's Dark Ages.
This Celtic cross dates
from the 900's.

Irish rebellion

During the late 1800's, some Irish people began to demand *home rule,* under which Ireland, as part of the United Kingdom, would have had its own parliament for domestic affairs. The British Parliament twice rejected home rule bills.

Republicans—Irish people who wanted a completely independent Irish republic—rebelled against the British in Dublin on Easter Monday 1916. The Easter Rebellion was easily put down, but in 1918 the republicans won most of Ireland's seats in the British Parliament. Instead of going to London to take their seats, they formed the *Dáil Éireann* (House of Representatives) in Dublin, and in 1919 they declared all of Ireland an independent republic.

As a result, fighting broke out between British forces and the Irish Republican Army (IRA), as the rebels were called. In 1920, the British Parliament passed an act that divided Ireland into two separate countries—one consisting of six northern counties and the other of the rest of Ireland. The northern counties accepted the act and formed the state of Northern Ireland. The southern counties rejected the act and fought on. Finally, in 1921, the United Kingdom signed a treaty making southern Ireland a *dominion* (self-governing country) of the British Commonwealth called the Irish Free State. In 1949, all ties with the United Kingdom were finally cut, and Ireland became an independent republic.

Bitterness between Catholics and Protestants in Northern Ireland erupted into violence in the late 1960's. In 1994, paramilitaries on both sides declared a cease-fire, but the IRA resumed terrorist bombings in 1996. In 1998, peace talks on Northern Ireland concluded in an agreement that promised an end to the conflict in the troubled region. In 1999, the United Kingdom ended direct rule of Northern Ireland. The Republic of Ireland, in turn, gave up its claim to Northern Ireland.

THE IRISH CHARACTER

The Irish are generally considered to be warm, friendly, talkative people. Legend connects their "gift of gab" with the Blarney Stone—a block of limestone in Blarney Castle, near Cork. According to the legend, anyone who kisses the stone receives the gift of expressive, convincing speech. Today, the word *blarney* has come to mean flattering or coaxing talk.

Literature and music

The traditional Irish way with words also extends to writing, and Ireland has produced many great writers over the centuries. In the late 1800's, a number of authors set out to create a body of literature in the English language that would express the Irish experience and tradition.

During this period of Irish literary revival, the poet William Butler Yeats and the playwright Lady Gregory founded Dublin's Abbey Theatre. The theater produced dramas by Sean O'Casey, John Synge, Yeats, and other gifted Irish playwrights. Some of the plays dealt with Irish peasant life, while others, such as O'Casey's *The Shadow of a Gunman* (1923) and *Juno and the Paycock* (1924), reflected the political upheaval raging in Ireland during the 1920's.

Music is also a vital part of Irish life, especially the folk music played on the harp and the bagpipes. Irish folk songs can as easily tell the tale of a struggling farmer or a lost love as they can relate a lot of comical "blarney" about a fast-talking Irishman.

Everyday life

While most of the people in Irish cities and towns live in houses, apartment living is on the rise. Many apartments are located in buildings above grocery stores and other shops.

At one time, large numbers of young people lived with their parents and remained single until they were well over 30. This was largely because farmland and jobs were scarce, and few young people could afford to marry and raise families. Today, with more job opportunities having opened up in urban areas, many young people leave home and marry earlier.

A strolling musician entertains at the Whitegate Pony Races in County Clare. Folk songs and traditional music are popular among the Irish people.

A grocer in Athlone, a town in County Westmeath, proudly displays his wares. Many shopkeepers in towns and cities live in apartments above their shops.

A poetry reading group gathers in a weekly meeting above a pub (public house) in Dublin. Pubs play an important role in Ireland's cultural and social life.

A young man in Connemara, County Galway, loads peat on his donkey's back just as his ancestors did. Peat, which consists of decayed plant material—and is an early stage in the formation of coal—is used as a household fuel.

Modern houses have replaced most of the traditional thatch-roofed cottages that once dotted Ireland's countryside, but people in rural areas live much as their ancestors did. For instance, peat, which is used as fuel in Ireland, is still cut by hand and transported in baskets carried by donkeys.

In their leisure time, many Irish people enjoy sports, and horse racing is one of their favorites. Races are held throughout the year on dozens of race tracks around the country. The most famous events include the Irish Derby, held at Kildare in late June, and the Irish Grand National, held near Dublin on the Monday after Easter. Thousands of visitors also enjoy the Royal Dublin Society Horse Show held in Dublin each August.

Ireland's favorite team sports are soccer; *Gaelic football,* which resembles soccer; and *hurling,* which is similar to field hockey. Other popular team sports include *camogie,* which is also similar to field hockey and is played by women; cricket; and Rugby football. Many Irish people also enjoy boxing, fishing, and golf.

ISRAEL

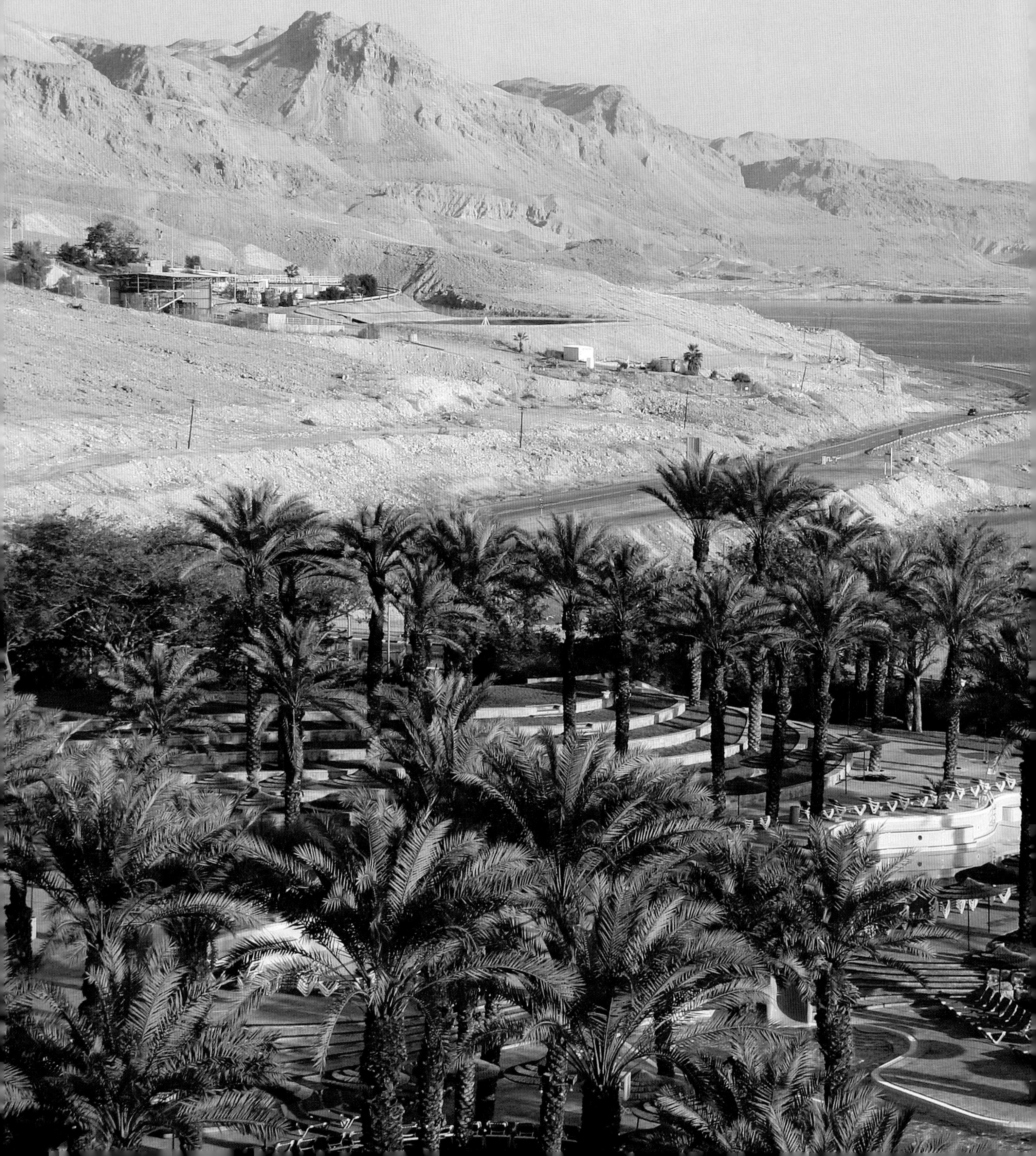

Israel is a small, historic country on the eastern shore of the Mediterranean Sea. The nation was created in 1948, but Israel's roots go back thousands of years. Most of the nation's people are Jews, and Jews all over the world consider Israel their spiritual homeland. But the people of Israel belong to a variety of ethnic groups, and Palestinians and other Arabs make up an important ethnic minority. The land of Israel and, in particular, the city of Jerusalem are sacred places to three major world religions: Judaism, Christianity, and Islam.

The Hebrew people settled in the area, which was then called Canaan, about 4,000 years ago. Part of the Hebrews' land came to be named Judah, and the people came to be called Jews. Most Jews fled their homeland after A.D. 135, when the Romans put down Jewish revolts against their rule. The Romans began to call this region by the word that became *Palestine* in English. The Jews were unable to re-create their nation until 1948, when modern Israel was created as a homeland for the Jewish people after World War II.

Israel soon became a modern democratic state with a developed economy. However, the creation of Israel caused great problems in the Middle East. Most Arabs opposed its founding, and the result has been a number of wars between Israel and Arab nations, as well as much terrorist violence against Israel. In some of the conflicts, Israel has gained territory.

Hope for an end to the fighting dawned in 1993, when Israel and the Palestine Liberation Organization (PLO) signed a peace accord. But despite peace efforts, violence continued during the 1990's and early 2000's. Disagreements about Jewish settlement in the West Bank and the Gaza Strip have led to continued distrust between the Israelis and Palestinians.

Israel has a pleasant climate, with hot, dry summers and cool, mild winters. The country has four major land regions: the coastal plain, the Judeo-Galilean Highlands, the Rift Valley, and the Negev Desert. The narrow coastal plain is home to most of the nation's people, industry, and farmland. In this region lie two of Israel's three largest cities, Haifa and Tel Aviv, and the fertile plains of Esdraelon and Sharon.

East of the coastal plain are highlands and hills that run from Galilee in the north to the Negev Desert in the south. Many of Israel's Arabs live in Galilee. Both Nazareth, the largest Arab center, and Jerusalem, the country's largest city, are in the highlands. The wedge-shaped Negev is an arid area of flatlands and mountains, but irrigation projects have turned sections of this desert into fertile farmland.

East of the highlands and desert lies the Rift Valley, which extends into Africa. The edges of the valley are steep, but the floor is flat and low. In the northern Rift Valley, the River Jordan flows through the Sea of Galilee to the Dead Sea. The shore of the Dead Sea is the lowest land area on Earth.

ISRAEL TODAY

Although Israel is a small country with few resources, its people have created a modern, democratic, economically developed nation. Israel has no written constitution. The government follows "basic laws" passed by its parliament, the Knesset, which makes laws and helps form national policy. The 120 members of the Knesset are elected for four years.

All Israeli citizens 18 years old or older may vote, but Israeli voters cast ballots for party lists, rather than for individual candidates. A party list includes all the candidates from 1 to 120—of a particular political party. Israel has many different parties reflecting many different views, but three parties dominate national elections—the Labor Party, Kadima, and the Likud bloc.

The moderate Labor Party supports some government control of the economy, with limited free enterprise. It also favors negotiating with neighboring Arab states. It sometimes forms alliances with Kadima, another moderate party. The Likud bloc, on the other hand, supports limited government control of the economy and takes a hard line toward the Arab states. The bloc is actually an alliance of a number of smaller parties.

The outcome of an Israeli election is determined by the percentage of the vote given to each party list. If, for example, a particular party list received one-third of the vote, that party would control one-third of the Knesset, or 40 seats.

Usually, the leader of the party that controls the most seats becomes prime minister. The prime minister heads the Cabinet, whose members head government departments and serve as Israel's top policymaking body. Israel also has a president elected by the Knesset, but the office is largely ceremonial.

The prime minister must have the support of a majority of the Knesset to stay in power. If the Labor Party or Likud bloc has too few seats to form a majority, it usually seeks support from the minor religious or special-interest parties in Israel. In this way, these small parties have considerable power.

FACTS

Official name:	Medinat Yisra'el (State of Israel)
Capital:	Jerusalem
Terrain:	Negev Desert in the south; low coastal plain; central mountains; Jordan Rift Valley
Area:	8,522 mi² (22,072 km²)
Climate:	Temperate; hot and dry in southern and eastern desert areas
Main river:	Jordan
Highest elevation:	Mount Meron, 3,963 ft (1,208 m)
Lowest elevation:	Shore of the Dead Sea, about 1,381 ft (421 m) below sea level
Form of government:	Democratic republic
Head of state:	President
Head of government:	Prime minister
Administrative areas:	6 mehozot (districts)
Legislature:	Knesset (Parliament) with 120 members serving four-year terms
Court system:	Supreme Court
Armed forces:	176,500 troops
National holiday:	Independence Day - May 14 (1948)
Estimated 2010 population:	7,279,000
Population density:	854 persons per mi² (330 per km²)
Population distribution:	92% urban, 8% rural
Life expectancy in years:	Male, 79; female, 82
Doctors per 1,000 people:	3.7
Birth rate per 1,000:	21
Death rate per 1,000:	5
Infant mortality:	4 deaths per 1,000 live births
Age structure:	0-14: 28%; 15-64: 62%; 65 and over: 10%
Internet users per 100 people:	29
Internet code:	.il
Languages spoken:	Hebrew (official), Arabic (official for Arab minority), English
Religions:	Jewish 76.4%, Muslim 16%, Christian 2.1%, Druze 1.6%, other 3.9%
Currency:	New shekel
Gross domestic product (GDP) in 2008:	$199.50 billion U.S.
Real annual growth rate (2008):	3.9%
GDP per capita (2008):	$27,157 U.S.
Goods exported:	Chemicals, cut diamonds, electronics, food, machinery, pharmaceuticals
Goods imported:	Chemicals, crude oil and petroleum products, electronics, food, iron and steel, machinery, motor vehicles, rough diamonds
Trading partners:	Belgium, China, France, Germany, Hong Kong, Italy, United Kingdom, United States

Education is a high priority in Israel. Education is free, and children between the ages of 5 and 16 are required to attend school.

The people of Israel enjoy a fairly high standard of living and a low level of unemployment. Israel began as a poor country, but its people have established industries, drained swamps, and irrigated the desert.

Today, Israel has a largely service-based economy. About 75 percent of Israeli workers are employed in service industries, such as community work, trade, or tourism. Many of these workers are employed by the government or work in businesses owned by the government.

A significant number of Israel's business firms are government owned. The *Histadrut* (General Federation of Labor), a powerful organization of trade unions, also owns a number of businesses, farms, and industries. But most of the nation's businesses are privately owned.

About one-sixth of Israeli workers are employed in manufacturing. They make such goods as chemical products, computer and electronic equipment, fertilizer, paper, plastics, processed foods, and textiles and clothing. The government-owned plants produce equipment needed by Israel's armed forces to maintain military readiness.

Israelis have transformed their dry, unproductive land into farmland through irrigation. Using modern methods and machines, a small number of farmers now produce enough food to meet domestic demands and pay for any necessary food imports. Fruits, cotton, eggs, grains, poultry, and vegetables are the main food products.

Despite its forbidding name, the Dead Sea is a source of mineral wealth. Table salt and *potash,* a mineral used for fertilizer, are extracted from the water.

Israel is a small Middle Eastern country on the eastern shore of the Mediterranean Sea. Since it officially came into existence on May 14, 1948, it has fought several wars with its Arab neighbors.

Diamond cutters pursue their craft in Israel, where cutting imported diamonds is a major industry. Other Israeli industries produce such high-technology goods as electronic equipment and scientific instruments.

HISTORY

Twelve Hebrew tribes created the Kingdom of Israel about 3,000 years ago, but the kingdom later split in two. The Hebrews—who came to be called Jews—fell under the rule of a series of empires. Assyrians, Babylonians, Persians, and Greeks all conquered and ruled the Jews. The last of the conquerors were the Romans. In A.D. 135, Rome drove the Jews out of their capital of Jerusalem and renamed the area *Palestine*.

The move to establish the modern state of Israel began in the late 1800's, when Theodor Herzl, an Austrian, founded the Zionist movement. Zionists sought to establish a homeland for the world's Jews in Palestine.

In 1917, the United Kingdom issued the Balfour Declaration, stating its support for such a Jewish homeland. At the time, the United Kingdom was trying to win control of Palestine from the Ottoman Empire during World War I (1914-1918).

Palestine came under the administration of the United Kingdom after the war ended. Large numbers of Jewish immigrants fled to Palestine during the 1930's to escape persecution from Nazi Germany. Arabs living in the region became alarmed and rebelled against the British.

During World War II (1939-1945), millions of Jews were killed by the Nazis. After the war, demands for a Jewish state increased, and the United Kingdom turned the problem over to the United Nations (UN). The UN devised a plan to split Palestine into a Jewish nation and an Arab nation.

In 1948, the State of Israel was established under the leadership of David Ben-Gurion. But Arab nations refused to recognize Israel and attacked the new state. Israel quickly de-

Jewish refugees from Europe flocked to Palestine after World War II, despite efforts by the British to limit Jewish immigration to the area. The newcomers helped found the State of Israel in 1948.

feated the Arabs and gained control of about half the land planned for the new Arab state. About 150,000 Palestinian Arabs thus became part of Israel's population.

In 1967, Egypt sent troops into the Sinai and blocked an Israeli port. Israel launched an air strike and almost completely destroyed the air forces of Egypt and its allies, while its ground troops defeated their armies.

By the war's end, Israel had won control of Egypt's Sinai Peninsula and Gaza Strip, Syria's Golan Heights, and Jordan's West Bank. With this land came 1 million more Palestinian Arabs. The Palestine Liberation Organization (PLO) became especially active and launched attacks on Israelis.

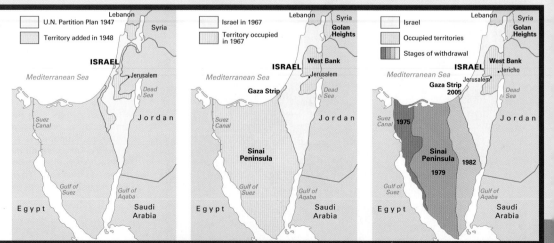

Israel's borders have varied since the country was established in 1948. When the UN plan to divide Palestine into Jewish and Arab states failed in 1948, Israel won a victory as well as additional territory in its war against surrounding Arab nations. In 1967, in only six days, Israeli forces took the Sinai Peninsula, Gaza Strip, Golan Heights, and West Bank. The 1979 peace treaty with Egypt resulted in staged withdrawal of Israeli troops from the Sinai. Israel holds the other territories and has established settlements in the West Bank.

1900-1700 B.C.	Hebrews (Israelites) settle in Canaan.
1600-1500 B.C.	Some Hebrews move to Egypt.
1200's B.C.	Moses leads Hebrews out of Egypt.
1000 B.C.	David forms unified Kingdom of Israel.
900's B.C.	King Solomon builds Temple.
922 B.C.	Nation splits into states of Israel and Judah.
722 or 721 B.C.	Assyrians conquer Israel.
587 or 586 B.C.	Babylonians conquer and exile Jews. Solomon's Temple is destroyed.
167 B.C.	Maccabeans lead Jewish revolt and establish kingdom of Judah.
63 B.C.	Romans occupy Judah (Judea).
A.D. 66	Romans put down Jewish revolt.
135	Romans drive Jews out of Jerusalem.
1800's	Zionist movement seeks to make Palestine a Jewish state.
1917	Balfour Declaration states British support for Jewish homeland.
1920	Palestine comes under British administration.
1930's	Jewish immigration sparks Arab violence.
1947	UN divides Palestine into Jewish state and Arab state.
1948	State of Israel created.
1948-49	Israel wins territory in the first Arab-Israeli war.
1956	Israel, the United Kingdom, and France attack Egypt in second Arab-Israeli war.
1964	Palestine Liberation Organization (PLO) founded.
1967	Israel wins control of Sinai, Golan Heights, and West Bank in Six-Day War.
1973	Israel fights Egypt and Syria in Yom Kippur War.
1978	Israel and Egypt agree to the Camp David Accords.
1979	Israel and Egypt sign peace treaty.
1982	Israel attacks PLO camps in Lebanon.
1985	Israeli troops leave most of Lebanon.
1987	Intifada begins.
1990	Police kill 21 Palestinian demonstrators.
1991	Iraq launches missile attacks on Israel.
1993	Israel and PLO sign peace accord for interim Palestinian self-rule.
1995	Israeli Prime Minister Yitzhak Rabin assassinated.
2000	Violence erupts between Palestinians and Israelis.
2005	Israel completely withdraws from the Gaza Strip and several West Bank settlements.
2006	The United Nations Security Council brokers a cease-fire between Israel and Hezbollah militants based in Lebanon.
2007	Israeli Prime Minister Ehud Olmert begins peace talks with Palestinian President Mahmoud Abbas.

Theodor Herzl
(1860-1904)

David Ben-Gurion
(1886-1973)

Yitzhak Rabin
(1922-1995)

The fourth Arab-Israeli war broke out in 1973. Egypt and Syria attacked Israel, and Israel pushed the Arab forces back. Tensions eased after the war, and in 1978, Egyptian President Anwar el-Sadat and Israeli Prime Minister Menachem Begin met with United States President Jimmy Carter. Their discussions resulted in the Camp David Accords, and a peace treaty was signed in 1979. Violence continued, however.

In 1993, Israeli Prime Minister Yitzhak Rabin and Yasir Arafat, chairman of the PLO, signed an accord in which Israel would cede administrative control of Palestinian land gained in the 1967 war. In 1994, Israel and Jordan signed a nonag-gression pact. But in 1995, Rabin was assassinated by an Israeli Jew who opposed the peace process. Following the assassination, Shimon Peres became prime minister. Benjamin Netanyahu, a critic of the Israel-PLO peace agreements, defeated Peres in 1996. Netanyahu's decision to resume construction of Israeli settlements in the West Bank was met with protests by Palestinians. In 1997, Israel agreed to withdraw Israeli troops from most of the West Bank city of Hebron. In 1998, Netanyahu, claiming that the PLO was not fulfilling its security commitments, suspended Israeli troop withdrawals.

In 2001, Ariel Sharon was elected prime minister of Israel. He was reelected in 2003. During his years in office, attacks by Palestinian militias and suicide bombers in Israel, the West Bank, and the Gaza Strip killed hundreds of Israelis. Israeli forces retaliated, killing more than 1,800 Palestinians. In 2002, Israel reoccupied much of the West Bank. In 2005, under Sharon's direction, all Jewish settlers were evacuated from the Gaza Strip and from some West Bank settlements.

In 2006, fighting between Israeli forces and Hezbollah militants based in Lebanon resulted in more than 1,000 deaths. The United Nations Security Council brokered a cease-fire.

Sharon suffered a stroke in January 2006 and was replaced as prime minister by Ehud Olmert. That same month, the radical Islamic organization Hamas won a landslide victory in Palestinian parliamentary elections. Israel negotiated a cease-fire with Palestinian militants, and Olmert began peace talks with Palestinian President Mahmoud Abbas in 2007. Hostilities resumed in 2008 and a new cease-fire was agreed upon in 2009, though some fighting continued. Olmert resigned as prime minister in 2008 over corruption charges. He was replaced in 2009 by former Prime Minister Benjamin Netanyahu.

PEOPLE

When Israel was established in 1948, about 800,000 people lived there. Today, the population numbers more than 7 million.

About 1.8 million Jews migrated to Israel in its first 40 years, including many who fled from persecution in their home countries. The Israeli government allows any Jew, with a few minor exceptions, to settle in Israel.

Heritage

Today, about 75 percent of Israel's people are Jewish. But while they share a common spiritual and historical heritage, they differ in their religious observance. About one-fifth of the Jewish population maintains strict observance of the principles of Judaism, and about half of the Jewish people observe some, but not all, of these principles. The rest of Israel's Jewish population is *secular* (nonreligious). These different groups disagree on the role Jewish religion should play in the government of Israel.

Because Israel's Jews came from many different countries, they belong to many different ethnic groups—each with its own cultural, political, and historical background. Israel's Jewish population falls into two major cultural groups—the *Ashkenazim* and the *Sephardim*.

A Palestinian Arab tends her sheep in the Negev Desert. The Arabs live mainly in their own rural villages or in separate neighborhoods in Israeli cities.

The Ashkenazim came to Israel from Europe or North America and are descended from members of the Jewish communities of central and eastern Europe. Many Ashkenazim speak Yiddish, a Germanic language. The Sephardim, on the other hand, came from countries in the Middle East and the Mediterranean.

A divided society

Today, most Israeli Jews are Sephardim, but in the early years of the nation, the Ashkenazim dominated. Many of these early Ashkenazim were skilled tradespeople and professionals who made great contributions to Israel's development. As a result, Israel's political, educational, and economic systems are much like those of Western nations. The Sephardim have had to adapt to this "foreign" society.

An Israeli soldier carries out a security check on an Arab resident of Jerusalem following rioting in the West Bank. Many Palestinian Arabs oppose Israeli rule.

Another group that has had a difficult time adapting to Israeli society is the Arab population. Arabs are by far the largest minority in Israel, making up nearly all the non-Jewish population. Most Arabs are Palestinians whose families remained in Israel after 1948.

The Jews and Arabs of Israel are often suspicious—and sometimes openly hostile—toward one another. Most live in separate areas, speak different languages, attend different schools, and follow different cultural traditions.

Israel has two official languages—Hebrew, the language spoken by most of the Jews, and Arabic, spoken mainly by the Arabs. The nation has two school systems—a Jewish system, in which instruction is given in Hebrew, and an Arab system, in which instruction is given in Arabic. The government recognizes and funds both systems. Although Israel's government guarantees religious freedom and allows members of all faiths to observe their Sabbaths and holy days, tension between the Jews and Arabs remains.

Urban life

More than 90 percent of Israel's people live in urban areas. About 20 percent of the nation's people live in the country's three largest cities—Jerusalem, Tel Aviv, and Haifa. Jerusalem is the capital and largest city. Tel Aviv is a center of industry and commerce, and Haifa is Israel's main port city.

Many Israeli cities are built on the sites of ancient settlements where there are many historic buildings. But Israelis have also constructed large, modern sections with high-rise office buildings. Most urban Israelis live in modern apartments. They face the same problems as people in other rapidly growing urban areas, including traffic congestion, housing shortages, and pollution.

Although most Israelis wear Western-style clothing, their food and drink reflects their ethnic diversity. For example, traditional European Jewish dishes, such as chopped liver, chicken soup, and gefilte fish, are common. But so are traditional Middle Eastern foods such as *felafel*—small, deep-fried patties of ground chickpeas. However, all government buildings and most restaurants and hotels serve only *kosher* foods, which are prepared according to Jewish dietary laws.

Sephardic (mainly eastern) Jews, like this Yemeni bride, follow customs different from those of Ashkenazi (European) Jews.

Israelis relax on a beach in Tel Aviv. The city lies on the eastern shore of the Mediterranean Sea. Tel Aviv is one of the most modern cities in the Middle East. It is the country's chief commercial, financial, and industrial center.

JERUSALEM

Jerusalem, Israel's capital city, is a holy city to Jews, Christians, and Muslims. This ancient city has been the site of much conflict through the ages.

When modern Israel was established in 1948, Tel Aviv was named its capital, and Jerusalem was to be an international city. But when Israel was attacked by neighboring Arab nations, Jerusalem became the scene of fierce fighting. By the end of 1948, Israeli soldiers held West Jerusalem, and Jordanian troops controlled East Jerusalem. The Israelis then declared West Jerusalem their capital city. However, most countries that had diplomatic relations with Israel refused to recognize West Jerusalem as the nation's capital.

When war between Israel and Arab nations broke out again in 1967, Israel captured East Jerusalem and announced it would keep all of Jerusalem. In 1980, the Israeli government officially named the city of Jerusalem as its capital.

A sacred city

Israelis felt deeply that Jerusalem should be their capital because of its long history as a holy city to Jews. Jerusalem was their political and religious center in Biblical times. About 1000 B.C., King David captured the town from a people called the Jebusites and made it the capital of the Kingdom of Israel.

King David's son, King Solomon, built a magnificent house of worship—the first Temple of the Jews—in the city. Over hundreds of years, the Temple was captured, recaptured, destroyed, and rebuilt several times. Finally, to put down a Jewish revolt, the Romans burned the Temple in A.D. 70. Today, the Western Wall, also called the Wailing Wall, is all that remains of the Jews' holy Temple. The wall, which is 160 feet (49 meters) long, has long been a symbol of Jewish faith and unity. When Jerusalem was split after the war of 1948, many Jews were bitterly disappointed because the wall lay in Arab-controlled East Jerusalem.

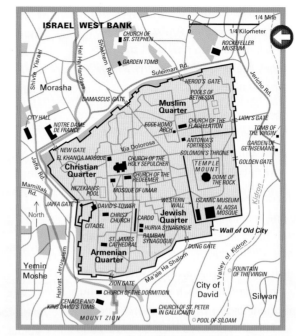

The Old City, in eastern Jerusalem, contains many sites that are sacred to Christianity, Islam, and Judaism. The city is divided into the Armenian, Christian, Jewish, and Muslim quarters.

Jews pray at the Western Wall in the Old City. The wall is all that remains of the Second Temple destroyed by the Romans in A.D. 70. The wall is also sometimes called the Wailing Wall, for the sorrowful prayers said there.

Christians also regard Jerusalem as a holy city because Jesus Christ was crucified there. The Church of the Holy Sepulcher is believed to stand on the hill of Calvary, or Golgotha, where Jesus Christ was crucified and buried. The Via Dolorosa (Way of Sorrows) is the route that Christians believe Jesus walked, carrying the cross. Many other events in the life of Christ also took place in Jerusalem.

To Muslims, Jerusalem is holy because they believe that Muhammad, the founder of their religion, rose to heaven from the city. A beautiful golden-domed monument called the Dome of the Rock marks the site. According to Muslim belief, Muhammad rose with the angel Gabriel and spoke to God, then returned to spread the religion of Islam. A 1994 peace pact with Israel gave Jordan a role in administering the Dome of the Rock. This pact was denounced by Palestinians, who felt it weakened their claim to part of the city.

The skyline of Jerusalem's Old City is dominated by the golden Dome of the Rock, an important Muslim shrine. Modern skyscrapers rise in the distance.

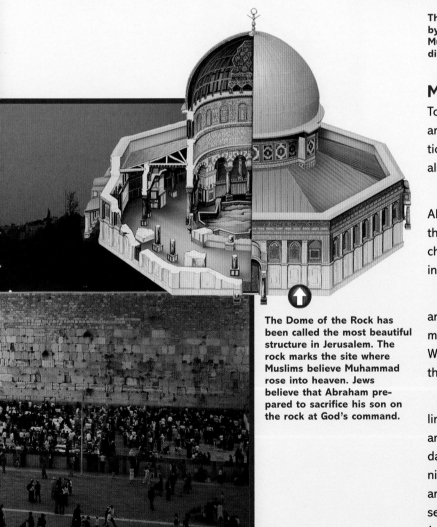

The Dome of the Rock has been called the most beautiful structure in Jerusalem. The rock marks the site where Muslims believe Muhammad rose into heaven. Jews believe that Abraham prepared to sacrifice his son on the rock at God's command.

Modern-day Jerusalem

Today, about two-thirds of the people of Jerusalem are Jews. They live in West Jerusalem, the newer section of the city, where some Christians and Muslims also live.

Almost all the people of East Jerusalem are Arabs. About four-fifths of them are Muslims, and most of the others are members of various eastern Christian churches. East Jerusalem includes the oldest district in all Jerusalem—the Old City.

The Old City lies on the site of ancient Jerusalem and is surrounded by stone walls almost 40 feet (12 meters) high and 2-1/2 miles (4 kilometers) long. Within these walls lie the Western Wall, the Church of the Holy Sepulcher, and the Dome of the Rock.

Jerusalem is a city of three Sabbaths—Friday (Muslim), Saturday (Jewish), and Sunday (Christian). Stores and businesses may be closed on any of these three days. After the Jewish Sabbath begins on Friday night, a large portion of West Jerusalem closes down, and there is no public transportation. Most Jews observe the Sabbath, but many others go to East Jerusalem where cafes and other places of entertainment are open.

LIFE IN A KIBBUTZ

For many people throughout the world, the *kibbutz* symbolizes the nation of Israel. In Israel, a kibbutz is a community in which no one owns private property. All property belongs to the kibbutz, and the members work for the kibbutz. In return, the kibbutz meets all the needs of the members and their families. It provides food, housing, education, child care, and medical care.

The first kibbutz was founded in what was then Palestine in 1909. Two women and 10 men from Poland established a collective settlement called Deganya on the Sea of Galilee. Like other early kibbutzim, this settlement was an agricultural community. Today, more than 250 kibbutzim are found in Israel, and most own factories as well as farmland.

Kibbutzim range in size from about 50 to 1,000 members, but a typical kibbutz has about 250 members. Israelis who wish to join a kibbutz usually work on the kibbutz for a year as a candidate for membership. Members of the kibbutz then vote on whether to accept the candidate.

All kibbutz members have an equal say in how the community is run. Many kibbutz members have regular daily jobs, while others are assigned to a variety of jobs by a work committee. Traditionally, members received no pay for their work but received all the goods and services they needed. Since the 1980's, however, some kibbutzim have begun paying wages or charging members for food and utilities.

In some kibbutzim, children sleep in their parents' homes. In other kibbutzim, they sleep in children's houses. In either case, children spend most of the day with their peers. From kindergarten on, their education emphasizes the importance of cooperation. Children are assigned duties, and by high school they spend one day each week working in the kibbutz.

Mothers of infants visit their children frequently during the day, and both parents join their children for a time after work. In this way, parents and children form close ties even when they do not live together.

Israelis who have been born and raised on kibbutzim are known for being high achievers. Many kibbutz mem-

Crates of peaches are packed for export at an Israeli farm. Most farms are organized as either moshavim or kibbutzim. On moshavim, the land is privately owned; on kibbutzim, the land belongs to the entire community.

Bringing in the harvest, a young Israeli woman hoists large containers that soon will be filled with produce. Because life in a kibbutz emphasizes equality, men and women are expected to share all kinds of work.

An aerial view of a kibbutz on the shores of the Sea of Galilee shows its green, well-tended fields. Although they had to struggle to feed their members in the early years, kibbutzim are now modern, well-equipped farms that help feed the entire nation.

Volunteer workers take a break from their labor on a kibbutz. Young people from many countries take working vacations in such settlements in order to experience communal life.

bers have served in the Knesset or the Cabinet, and others have distinguished themselves in the army. Some people believe these achievements are due to the superior education and health care received on kibbutzim. Also, kibbutz members value the group more than the individual, and they learn to work well in many different group situations.

Moshavim or *moshavim shitufi* are two other kinds of rural communities. In a moshav, each family works its own land separately and lives in its own house, but the entire moshav owns and shares the large pieces of farm equipment. The moshav purchases essential supplies, such as seeds, and markets all the crops. Moshav members elect assemblies to supervise the moshav.

A moshav shitufi is like a moshav in some ways and a kibbutz in others. As in a moshav, members of a moshav shitufi have their own family households. As in a kibbutz, the members work collectively on the land, and profits are shared.

Together, the kibbutzim and moshavim help produce almost all the food Israel needs.

ITALY

The country of Italy sits on a boot-shaped peninsula. It is a land of mountain ranges, rolling hills, and bustling cities—and a people of abundant spirit, energy, and optimism.

Italy is known for its rich cultural heritage and natural beauty. Its cities have spectacular churches and large central plazas. Their museums contain some of the world's best-known art. The countryside has warm, sandy beaches; high, glacier-topped mountain peaks; and rolling hills covered with green fields and vineyards.

Italy got its name from the ancient Romans. The Romans called the southern part of the peninsula *Italia,* meaning *land of oxen* or *grazing land.*

For hundreds of years, Italy has strongly influenced the course of Western civilization. One of the world's greatest empires, the Roman Empire, began with a small community of shepherds on the hillsides of central Italy. In the early A.D. 1300's, the city-states of Italy were the birthplace of the Renaissance, a great cultural movement during which many European scholars and artists studied the learning and art of ancient Greece and Rome.

Everywhere in Italy, there are reminders of the glories of its ancient past. Many monuments built by the ancient Romans still stand. In the city of Rome, art museums display the masterpieces of such Renaissance artists and sculptors as Michelangelo, Leonardo da Vinci, and Botticelli.

Yet for the Italians, time marches on. A six-lane highway now encircles the Colosseum, ancient Rome's greatest architectural wonder. Today, the splendid Roman baths provide a background for operas. In the footsteps of their ancestors, the Italians have pressed on, creating one of the most highly industrialized nations in Europe.

The country boasts several world-famous cities. Rome, the capital and largest city of Italy, was the center of the Roman Empire 2,000 years ago. Florence was the home of many artists of the Renaissance. Venice, with its intricate canal system, attracts tourists from all over the world.

Today, more than two-thirds of Italy's people live in cities and towns, and industrial plants have become as much a part of the Tuscan countryside as the area's legendary olive groves. But even so, Italy has a timeless quality that seems to fill the heart of every visitor.

221

ITALY TODAY

Italy is an ancient land but a young nation. It became a united country in the relatively recent year of 1861, when numerous smaller kingdoms on the Italian Peninsula came together under the leadership of King Victor Emmanuel II.

In 1922, Benito Mussolini became premier of Italy, and by 1925, he ruled as a dictator. After Mussolini was overthrown by King Victor Emmanuel III in 1943, Italy returned briefly to monarchy. Then, in 1946, the Italian people chose to replace the monarchy with a new government. In 1948, a new constitution proclaiming "a democratic republic founded on work" went into effect.

An economic shift

After World War II (1939-1945), Italy shifted from an agriculture-based economy to an industry-based economy. The growth of industry was so rapid and dramatic that industrial production had more than doubled its prewar level by the 1960's.

The rapid changes in Italy's economy also brought many problems. The gap between the rich industrial north and the poorer southern regions increased. The Italian government tried to address this gap and to improve conditions for workers. But the economy faltered, and many Italians began to see their government as inefficient or corrupt.

Political and social change

After the Italian republic was formed, three political parties—the Christian Democrats, the Socialists, and the Communists—became the most powerful. For many years, the Christian Democrats ruled the country, usually in coalitions with smaller parties.

During the 1970's, the Christian Democrats began to cooperate more fully with the Communists in an effort to strengthen the nation's struggling economy. This cooperation was strongly opposed by many Christian Democrats, as well as by some of Italy's allies. In addition, a leftist terrorist group called the Red Brigades worked to spread fear and disruption throughout Italy. In 1978, the Red Brigades kidnapped and murdered Aldo Moro, a former prime minister. Reaction to the murder led to a renewed drive against the terrorists, and hundreds of people were arrested and convicted.

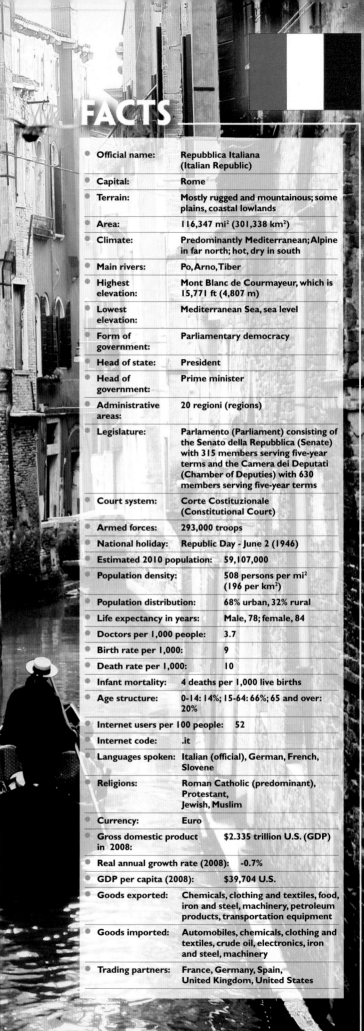

FACTS

Official name:	Repubblica Italiana (Italian Republic)
Capital:	Rome
Terrain:	Mostly rugged and mountainous; some plains, coastal lowlands
Area:	116,347 mi² (301,338 km²)
Climate:	Predominantly Mediterranean; Alpine in far north; hot, dry in south
Main rivers:	Po, Arno, Tiber
Highest elevation:	Mont Blanc de Courmayeur, which is 15,771 ft (4,807 m)
Lowest elevation:	Mediterranean Sea, sea level
Form of government:	Parliamentary democracy
Head of state:	President
Head of government:	Prime minister
Administrative areas:	20 regioni (regions)
Legislature:	Parlamento (Parliament) consisting of the Senato della Repubblica (Senate) with 315 members serving five-year terms and the Camera dei Deputati (Chamber of Deputies) with 630 members serving five-year terms
Court system:	Corte Costituzionale (Constitutional Court)
Armed forces:	293,000 troops
National holiday:	Republic Day - June 2 (1946)
Estimated 2010 population:	59,107,000
Population density:	508 persons per mi² (196 per km²)
Population distribution:	68% urban, 32% rural
Life expectancy in years:	Male, 78; female, 84
Doctors per 1,000 people:	3.7
Birth rate per 1,000:	9
Death rate per 1,000:	10
Infant mortality:	4 deaths per 1,000 live births
Age structure:	0-14: 14%; 15-64: 66%; 65 and over: 20%
Internet users per 100 people:	52
Internet code:	.it
Languages spoken:	Italian (official), German, French, Slovene
Religions:	Roman Catholic (predominant), Protestant, Jewish, Muslim
Currency:	Euro
Gross domestic product in 2008:	$2.335 trillion U.S. (GDP)
Real annual growth rate (2008):	-0.7%
GDP per capita (2008):	$39,704 U.S.
Goods exported:	Chemicals, clothing and textiles, food, iron and steel, machinery, petroleum products, transportation equipment
Goods imported:	Automobiles, chemicals, clothing and textiles, crude oil, electronics, iron and steel, machinery
Trading partners:	France, Germany, Spain, United Kingdom, United States

The Republic of Italy includes the boot-shaped Italian Peninsula as well as the islands of Sicily and Sardinia. The independent states of San Marino and Vatican City also lie within Italy's borders. Vatican City is located within the city of Rome.

Beginning in the late 1960's, the political influence of the Roman Catholic Church began to weaken in Italy. Divorce was legalized in 1970, and abortion, in 1978. In 1985, a 1929 agreement that had made Roman Catholicism the state religion was dissolved.

Organized crime has long been a problem in Italy. Organized crime activities range from international drug trafficking to corruption in awarding government construction contracts. In 1986, the government began the prosecution of hundreds of accused Mafia members in Palermo. Many were convicted, but the Mafia retained considerable influence.

Since the 1990's, Italy's main political parties have tended to group themselves into two opposing coalitions, one moderately conservative and the other, moderately liberal. Silvio Berlusconi, a wealthy busi-

nessman, served as prime minister for seven months in 1994. He led a conservative coalition to victory in a 2001 election and served as prime minister until 2006, when he lost a close election to businessman Romano Prodi.

Berlusconi became prime minister again in 2008. Dogged by numerous scandals and an economic crisis, he resigned in 2011. An economist, Mario Monti, was named the new prime minister in the hope that he would save Italy from economic collapse brought on by its huge national debt.

ENVIRONMENT

The mainland of Italy is a long, mountainous peninsula extending 708 miles (1,139 kilometers) from its northernmost point to the Mediterranean Sea. The Alps form the nation's northern border, separating Italy from France, Switzerland, Austria, and Slovenia.

Western Italy faces the Ligurian and Tyrrhenian seas, while the coast of southern Italy—the "sole" of its famous "boot"—opens to the Ionian Sea. Italy's east coast lies on the Adriatic Sea.

The landscape of Italy is varied and beautiful. Steep cliffs plunge into the sea on the southeast coast, while inland the rolling countryside features mile after mile of green fields and vineyards.

Italy is located in the Alpine belt, which cuts across southern Europe and Asia. It is one of the world's most geologically active regions. Along this belt lie the edges of the giant *plates* that make up Earth's crust. As the plates move slowly and continuously, the rocks at their edges are stretched and squeezed. This squeezing and stretching causes earthquakes and sometimes volcanic eruptions. In 2009, nearly 300 people were killed and more than 65,000 others were left homeless when an earthquake struck the Abruzzo region.

Italy has the only active volcanoes in Europe. Stromboli, in the Tyrrhenian Sea, is constantly active, but violent eruptions are rare because the lava flows freely, instead of building up internal pressure. Mount Etna, on the island of Sicily, has erupted at least 260 times since its first recorded eruption in about 700 B.C. Vesuvius, rising on the Bay of Naples, is probably the most famous volcano in the world. In A.D. 79, Vesuvius erupted and destroyed the ancient city of Pompeii.

A mountainous land

Most of Italy consists of mountains and hilly regions. The Alpine Slope region, which stretches across the northernmost part of Italy, includes huge mountains and deep valleys. The lower slopes are lined with forests. The high mountaintops consist of barren rocks and glaciers.

The Apennine Mountains extend almost the entire length of Italy. The northern Apennines are lush and green with some of the largest forests in the country and much pastureland. The central range supports farmland and grasslands. The southern Apennines are made up mainly of plateaus and high mountains, with few natural resources.

The columns of an ancient temple rise in a meadow at Paestum. Paestum was founded about 600 B.C. by Greek settlers, who called it Poseidonia. It became a Roman colony, and received its present name, in 273 B.C.

Cypress trees line a gravel road in the hill country of Tuscany. According to legend, Cupid's arrows were made of cypress wood. Cupid was the Roman god of love, and anyone shot by his arrows supposedly fell in love.

Scaligeri Castle, once a fortress of the rulers of Verona, stands on a narrow peninsula that juts out into Lake Garda in the south-central part of the Alpine Slope.

The landscape of Italy varies widely from region to region. In the north, soaring mountains give way to broad plains. In the southeastern part of the country, high plateaus end in steep cliffs that plunge into the Mediterranean Sea.

Fertile plains

Between the Alps and the Apennines lie the vast plains of the Po Valley, Italy's richest and most modern agricultural region. Waters from the Po River, the country's largest waterway, have created this fertile farmland.

The Western Uplands and Plains are second only to the Po Valley in agricultural importance. The northern part of the region includes the rich hill country of Tuscany and Umbria, where grain crops flourish and livestock graze. In the southern part of the region, the warm climate is ideal for growing apricots, cherries, lemons, peaches, vegetables, and wine grapes.

Apulia and the Southeastern Plains—a semiarid limestone plateau surrounded by fertile plains—form the "heel" of Italy's "boot." This region produces most of Italy's olive oil. Water for the area's olive trees comes from the Sele River by way of the 143-mile (230-kilometer) Apulian Aqueduct, built between 1905 and 1928.

Island regions

Like the mainland of Italy, the island of Sicily has a varied landscape of mountains and plains. Mount Etna dominates the northeastern side of the island. Severe soil erosion, caused in part by the clearing of forests, has made agriculture difficult on Sicily.

Sardinia, with a landscape consisting of mountains and plateaus, lacks good farmland. Artichokes, cereals, and grapes are grown only on the island's narrow coastal plains.

EARLY HISTORY

About 800 B.C., a people known as the Etruscans arrived in the coastal plains of northern Italy between the Arno and Tiber rivers. The Etruscans had the most advanced civilization in Italy, and in about 600 B.C., they took control of Rome and other towns in Latium, a region south of Rome.

Under the Etruscans, Rome grew from a community of farmers and shepherds to a wealthy trading center. In less than 100 years, the Romans became so powerful that they were able to drive out the Etruscans.

The Roman Republic

Over time, a republican government began to develop in Rome. Meanwhile, Rome was extending its control to cover the rest of the Italian Peninsula. Protection and limited Roman citizenship were offered to the people who were defeated by the Romans. In return, the Romans took soldiers and supplies from the conquered cities.

Marching on to victory in city after city, the Roman army soon took to the seas to conquer distant territories. Their triumphs marked the beginning of one of the world's mightiest empires.

The building of an empire

In the three Punic Wars fought between 264 B.C. and 146 B.C., Rome defeated the Carthaginians of North Africa and gained control of the Mediterranean Sea. Later, the Roman army conquered Greece and Macedonia.

The army's military conquests brought great riches to the Roman Empire, including tax revenues from conquered countries and looted property seized in wars. But the large numbers of slaves brought from conquered lands to Rome to work on the plantations created unemployment among the Romans and drove out the small farmers. As the gap widened between the rich and the poor, discontent grew among the Roman people.

Soon Rome was involved in great political turmoil. Conflicts among its leaders resulted in a se-

ries of civil wars. In 60 B.C., a three-man political alliance called the First Triumvirate was formed by Pompey, Julius Caesar, and Marcus Crassus.

When Caesar conquered Gaul, other Roman leaders grew fearful of his power and ambition, and they ordered him to give up his command. Instead, Caesar left Gaul and invaded Italy in 49 B.C. By 45 B.C., Caesar had defeated his political enemies, and he became sole ruler of the Roman world.

A year later, Caesar was assassinated by a group of aristocrats who hoped to revive the Roman Republic. After another civil war, Caesar's adopted son and heir, Octavian, became the first Roman emperor. He took the name *Augustus,* meaning *exalted.*

An era of peace

Under the reign of Augustus, the Roman Empire entered a period of stability and prosperity. The period, known as the *Pax Romana* (Roman peace), lasted about 200 years. But after Augustus died in A.D. 14, a series of *dynasties* (rulers of the same family) governed the empire. The central government in Rome could no longer hold the empire together, and provincial governors and army commanders began to declare themselves emperors.

Diocletian, a general who was named emperor by his troops in 284, divided the empire into east and west sections, each ruled by a different emperor. In 395, the empire was split into the West Roman Empire and the East Roman Empire.

The Colosseum, in Rome, ranks among the finest examples of Roman architecture and engineering, even though it survives only as a ruin. The Colosseum was ancient Rome's largest arena.

The Pantheon is a well-preserved ancient temple in the center of Rome. It was completed about A.D. 126. Since 1885, it has served as a national shrine and burial place for Italian national heroes.

TIMELINE

Julius Caesar (100?-44) B.C.)

Augustus Caesar (63 B.C.-A.D. 14)

Hadrian (A.D. 76-138)

The West Roman Empire soon proved unable to withstand the attacks of the Germanic peoples from the north. In 476, the Germanic chieftain Odoacer forced Romulus Augustulus, the last ruler of the West Roman Empire, from the throne.

By the mid-500's, the Christian popes who now ruled Rome had acquired great influence in religious and political matters throughout the Italian Peninsula. With the help of the Frankish king, Pepin the Short, and his son, Charlemagne, the popes defeated the Lombards. In 754, they established political rule in central Italy over what became known as the Papal States.

By A.D. 117, the Roman Empire extended over half of Europe, much of the Middle East, and the north coast of Africa. The empire controlled territory that completely enclosed the Mediterranean Sea.

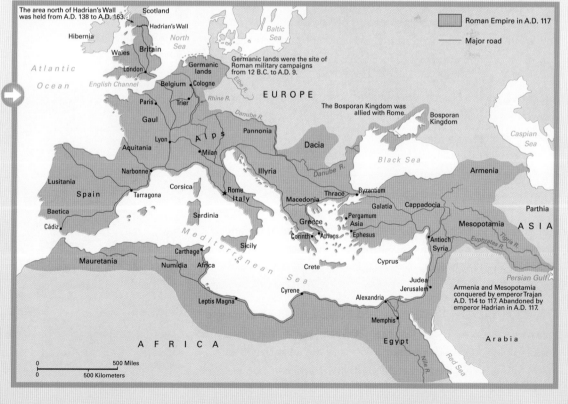

MODERN HISTORY

During the 1000's, the cities of Italy were ruled by what was later known as the Holy Roman Empire. Because the emperors lived in Germany, they had little direct power over the Italian lands. As a result, some Italian cities, such as Florence, Genoa, Milan, Pisa, and Venice, gradually developed into nearly independent *city-states*.

The Renaissance

Cultural life in the city-states helped encourage the growth of the cultural movement known as the Renaissance. Under the sponsorship of the Italian cities, painting, sculpture, and architecture reached great heights.

The word *Renaissance* comes from the Latin word *rinascere* and refers to the act of being reborn. Renaissance scholars and artists wanted to recapture the spirit of the ancient Greek and Roman cultures in their own artistic, literary, and philosophical works. The Renaissance thus represented a rebirth of these ancient cultures.

Some Renaissance philosophers—known as *humanists*—blended a concern for the history and actions of human beings with religious concerns. The humanists studied languages, literature, history, and ethics, believing that these subjects would help them better understand the problems of humanity. They used the culture of the ancient Greeks and Romans as a model for how they should conduct their lives.

Renaissance thinking soon spread to other European countries, and Italian styles influenced nearly every area of European activity. Italy soon became attractive to foreign conquerors. King Charles VIII of France marched into Italy in 1494, and the city-states could not hold back the French army. Charles soon withdrew, but he had shown that the cities of Italy could be conquered because they were not united.

Invaders and conquerors

In 1519, Charles I of Spain, a member of the Habsburg family, became Emperor Charles V of the Holy Roman Empire. His troops looted Rome in 1527, and by 1559 they had seized Milan and Sicily from France. Then, as Spanish influence weakened, control of Italy passed from the Spanish Habsburgs to the Austrian Habsburgs.

In 1796, the French ruler Napoleon Bonaparte led his army into Italy. French control lasted less than 20 years, but in that time, Napoleon introduced democratic reforms to the Italian people. Italy's former rulers were reestablished after Napoleon's defeat.

The huge dome of the Cathedral of Santa Maria del Fiore dominates the skyline of Florence. The dome's interior is decorated by *Last Judgment,* a fresco (wall painting) by Giorgio Vasari and Federico Zuccari. The cathedral, begun in 1296, is one of the finest examples of architecture of its period in Tuscany.

Thirteen towers built during the Middle Ages rise above the rooftops of San Gimignano, a hill town in Tuscany. More than 70 such towers were built in San Gimignano during that period, some as defense towers.

1000	Italian cities begin growing into independent city-states.
1000's	Normans conquer southern Italy and Sicily, and later unite them to form the Kingdom of Two Sicilies.
1215	Frederick II becomes emperor of the Holy Roman Empire.
c. 1300	The Renaissance begins in Italy.
1309-1377	Pope moves the papacy to Avignon, France.
1494	King Charles VIII of France marches into Italy.
1519	Charles I of Spain becomes emperor of the Holy Roman Empire.
1521-1559	The forces of Spain and the Holy Roman Empire defeat France in a series of wars over the control of Italy.
1700's	Austrian Habsburgs control most of Italy.
1796	Napoleon Bonaparte drives Austrians from Italy and seizes the country.
1814-1815	The Congress of Vienna returns Italy to its former rulers after Napoleon is defeated.
1848-1849	Italian states revolt against Austrian rule; Austrians crush the rebellion.
1861	The Kingdom of Italy is formed.
1871	Rome becomes the capital of Italy.
1882	Italy becomes a part of the Triple Alliance, along with Austria-Hungary and Germany.
1911-1912	Italy occupies Libya.
1915-1918	Italy fights on the side of the Allies in World War I (1914-1918).
1922	Benito Mussolini becomes premier of Italy.
1925	Mussolini rules as dictator of Italy.
1929	Lateran Treaty establishes normal relations between the Roman Catholic Church and the Italian government.
1936	Mussolini and German dictator Adolf Hitler sign an agreement that outlines a common foreign policy.
1940	Italy enters World War II (1939-1945) on the side of Nazi Germany.
1943	Italy surrenders to the Allies.
1945	Mussolini is executed by anti-Fascist Italians.
1946	The Republic of Italy is established.
1947	Italy's new constitution is adopted.
1950	Italy becomes a founding member of the North Atlantic Treaty Organization (NATO).
1958	Italy helps establish the European Economic Community, one of the groups that became the basis for the European Union (EU).
1978	Terrorist Red Brigades kidnap and kill former Premier Aldo Moro.
1999	Italy and most other members of the European Union adopt a common currency called the euro.
2011	Italy's large public debt, together with those of Greece, Ireland, Portugal, and Spain, threaten the viability of the euro.

Lorenzo the Magnificent, banker and Renaissance art patron (1449-1492)

Leonardo da Vinci, artist and inventor (1452-1519)

Giuseppe Garibaldi, forged independent Italy (1807-1882)

In 1866, the northwestern region of Venetia also became part of Italy. Only Rome, which remained under the pope's control, and tiny San Marino were not part of the new kingdom. By 1870, the pope's territory was reduced to the boundaries of the Vatican, and Rome became the capital of Italy in 1871.

After World War I (1914-1918) ended, many workers were unemployed and the Italian people became greatly discontented with their government. Soon, a movement called *Fascism,* led by Benito Mussolini, gained power. The Fascists favored strict government control of labor and industry. By 1925, Mussolini ruled as dictator.

In 1940, Mussolini led Italy into World War II on the side of Nazi Germany. After suffering a series of crushing defeats, Italy surrendered in 1943, and Mussolini was executed by anti-Fascist Italians in 1945. Italy was briefly ruled by a king before it became a republic in 1946.

But the Italian people, impressed by the ideals of the French Revolution, had begun to dream of a united, independent Italy.

Birth of a nation

In 1848, the Italian kingdoms revolted against their rulers, but Austria crushed the revolutions in 1849. After many years of fighting, the Italians drove out the Austrians. In 1861, Victor Emmanuel II, supported by a nationwide vote, declared the formation of the Kingdom of Italy and became Italy's first king.

The unification of Italy began in 1859, as parts of the Italian Peninsula joined with the Kingdom of Sardinia. The expanded kingdom became the Kingdom of Italy in 1861. The dates on this map indicate when each region joined the kingdom.

Boundary of present-day Italy

Savoy (to France 1860)

Trentino 1919

Istria 1919 (most to Yugoslavia in 1947)

Lombardy 1859

Venetia 1866

Trieste

Kingdom of Sardinia

Parma 1860

Modena 1860

Nice (to France 1860)

San Marino

Tuscany 1860

Papal States 1860

Adriatic Sea

Corsica (to France 1768)

Papal States 1870

Rome

Pontecorvo

Benevento

Kingdom of Sardinia

Tyrrhenian Sea

Kingdom of the Two Sicilies 1860

Mediterranean Sea

ROME

The interior of the Colosseum was renovated in the 1990's to permit modern theatrical productions. In ancient Rome, gladiators met a bloody death or achieved glorious victory inside the Colosseum for the entertainment of Roman audiences.

Almost 3,000 years ago, a group of shepherds built a small village on a hill along the banks of the Tiber River. From these simple beginnings came the great city of Rome, the center of Western civilization for more than 2,000 years.

Today, Rome is one of the most beautiful and historic cities in the world. From its ancient monuments to its priceless works of art, its *piazzas* (squares) and open-air markets, the city is a treasure-trove of Western culture. Because of its long history, Rome is known as the *Eternal City*.

The city stands on about 20 hills, which include the famous seven hills on which ancient Rome was built—the Aventine, Caelian, Capitoline, Esquiline, Palatine, Quirinal, and Viminal hills. The Tiber River flows through the center of Rome to the Tyrrhenian Sea.

Ancient beginnings

According to legend, Rome was founded by twin brothers, Romulus and Remus, who were heirs to the throne of the ancient Italian city of Alba Longa. Soon after their birth, their uncle, who wanted to remain king, had Romulus and Remus put into a basket and thrown into the Tiber River. Legend says that a female wolf found the twins and nursed them. Then they were rescued by a shepherd, who, with his wife, raised the two boys.

When they were young men, Romulus and Remus set out to found a city of their own. To settle a quarrel about the site of the proposed city, they decided that whoever counted the most vultures in flight would select the site. Romulus saw 12 vultures, while Remus saw only 6.

But Remus, suspecting that Romulus had cheated, mocked his brother. For this act of disloyalty, Remus was killed. Romulus then named the new city after himself and became the city's first ruler.

The story of Romulus and Remus and its bloody ending are part of Roman mythology, but the violence upon which much of Rome was built

was quite real. The ancient Romans conquered lands through military conquests, seized the riches they found there, and took the conquered citizens as slaves.

And the killing was not confined to the battlefield. From A.D. 80 until the 400's, the Colosseum was the scene of bloody battles staged for the entertainment of the Roman people. The slaughter included combat between gladiators, who were trained warriors, as well as between men and wild animals. On the day the Colosseum opened, 5,000 wild animals were slaughtered in its arena.

But in spite of its cruelty, the Roman Empire gave the Western world the foundations of its modern governmental, legal, and military systems, as well as its language and arts. Present-day Rome abounds with reminders of its ancient citizens—not only of their flaws but also of their triumphs.

Historic sites

Not far from the Roman Forum, the seat of ancient Rome's government, stands the Forum of Trajan. At its center, a column 100 feet (30 meters) high bears a striking series of sculptures depicting scenes from Tra-

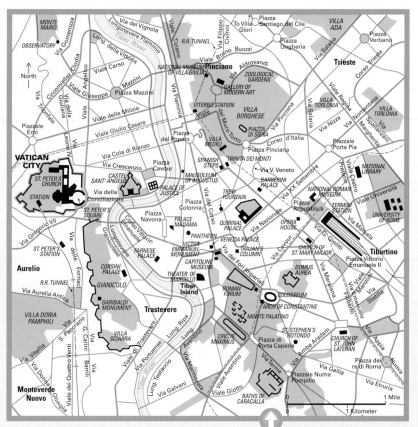

Romans and tourists alike relax on the Spanish Steps, one of the city's most popular meeting places. It is one of the many splendid sights of Rome, capital of Italy and once the center of Western civilization. Unlike most leading cities, Rome has little industry, and its economy depends chiefly on tourism.

Many ancient monuments and historical ruins keep alive the fascination of Rome's long and colorful history. The River Tiber winds through the center of the city.

The Piazza Navona follows the oval form of the ancient Circus of Diocletian on which it rests. Across from the Fountain of Neptune stands the Church of Sant' Agnese in Agone, partly rebuilt in the 1600's. During the 1700's, the center of the piazza was flooded as part of a local festival.

jan's wars. Nearby stand the Markets of Trajan, a large semicircle of three-storied shops. One of the shops has been rebuilt to show how it looked in ancient times.

Underneath the streets of Rome lie the *catacombs*—underground passages cut into the rocks by the early Christians in the 200's and 300's. During the period of Roman persecution, the Christians took refuge in the catacombs.

Visitors to Rome have two favorite meeting places—the Fountain of Trevi and the Spanish Steps. According to legend, a visitor who tosses a coin into the fountain is assured of a happy return to Rome someday.

The 200-year-old Spanish Steps rise gently upward to a square dominated by the Church of Trinità dei Monti. From the Spanish Steps, tourists can join the local people in the Roman pastime of *dolce far niente* (delightful idleness) and simply watch the world go by.

PEOPLE

Overall, the people of Italy have much in common, including their language, their religion, and a national character built on the importance of the family. Religious faith and family ties are sources of strength for the Italian people. In many ways, these basic values have helped the Italian people through the transformations their country experienced during the 1900's.

About 95 percent of the people who live in Italy today belong to the same Italian ethnic group. Often noted for their black hair, olive skin, and dark eyes, the Italian people are descendants of the Ligurian peoples, the original inhabitants of the peninsula.

As different groups settled the land, they created some regional variations that can still be seen throughout Italy. The Arabs and Normans who once occupied southern Italy and Sicily left their mark. People of Greek and Albanian heritage can be found in Sicily, as well as in Calabria. In Friuli-Venezia Giulia, near the Slovenian border, many people have Slavic characteristics.

The largest ethnic minority is the Germans, who live in the Trentino-Alto Adige region bordering Austria. German is the first language of many people in this region. Slovenes, another ethnic group, live in the Trieste area, along the border of Italy and Yugoslavia, and speak *Slovene,* a Slavic language. A number of ethnic French people live in the Valle D'Aosta region, near Italy's border with France and Switzerland. Slovene and *Ladin,* a language similar to the Romansch of the Swiss, are spoken in northern Venetia. Since the late 1900's, Moroccans and Albanians have made up two of the larger groups that have immigrated to Italy.

Language and dialects

Italian is the official language of Italy. An increasing number of Italians now choose standard Italian, which developed from a dialect spoken in Tuscany, over their regional dialect. The standardization of language in Italy today is due partly to television, books, and education, and even to travel and military service.

A woman in Rome hangs laundry outside her apartment window to dry. Most city dwellers live in concrete apartment buildings, and many buy, rather than rent, their apartments. A few wealthy people live in single-family homes.

A sidewalk cafe is a popular meeting place for a group of friends in a Tuscan town. Italians often meet for a friendly chat over a glass of wine or a dish of gelato (creamy Italian ice cream).

Members of a religious community enjoy a passeggiata (stroll) through the Italian countryside. Families generally enjoy a stroll in the early evening as a way of relaxing, getting some exercise, and visiting with friends.

A band of accordion players serenades a woman in Calabria, a region in southwestern Italy. The Italians have a great love of music and a rich musical heritage. The first operas were composed in Florence in the 1590's.

Nevertheless, many dialects are still spoken throughout the country. In the northern region, these include Emilian, Ligurian, Lombard, Piedmontese, and Venetian. In central Italy, some people speak Corsican, Roman, Tuscan, or Umbrian. And in the south, the dialects are Abruzzese, Apulian, Calabrian, Neapolitan, and Sicilian.

Religion

About 95 percent of the Italian people are Roman Catholic, and the Roman Catholic Church has had a tremendous influence on the country's political and social life. Although church influence weakened in the late 1900's, religion still plays an important role in the lives of many Italians. While only about 25 percent of the Italian people attend church regularly, most are baptized, married, and buried with Roman Catholic rites.

Local festivals celebrating religious holidays are held in Italy throughout the year. During the first week of January, many communities sponsor pageants re-creating the arrival of the Magi, the three wise men who followed the star of Bethlehem when Jesus Christ was born. On Good Friday in Taranto, a procession of hooded *penitents* (people sorry for their sins) makes a 14-hour journey around the city.

Ascension Day in Cocullo, in the Abruzzi region, is celebrated with a procession in which pilgrims and singers carry live snakes in their hands. Perhaps the most spectacular festival is the annual Carnival, held during the 10 days preceding Lent (the 40 weekdays between Ash Wednesday and Easter).

Family life

The maintaining of close family ties has always been an important tradition in Italy. The family has its strongest roots in the small farming communities that existed throughout the nation before the coming of industrialization.

Most Italians no longer live in the farms and villages of their ancestors. Hoping to find a better life, many people moved to the cities. But the family values that were passed down through the generations remain strong.

FOODS OF ITALY

To many people, Italian food means—simply—pizza! The word *pizza* means *pie* in Italian, and the dish was invented in Naples in the 1700's. *Pizza napoletana verace* (true Neopolitan pizza), which consists of tomatoes, garlic, olive oil, and oregano baked on dough over a wood fire, can still be purchased in the tiny street stalls of Naples—though electric ovens have largely replaced the wood fires.

But for all its popularity, pizza is just one of the many delicious creations served throughout Italy. The abundance of fresh fruits, vegetables, herbs, and spices grown in the rich soil of Italy provide the basis for *la cucina Italiana* (Italian cooking).

The Italian people seem to enjoy eating as much as they enjoy cooking. An Italian meal is a festive social occasion as well as a way to fill an empty stomach.

An Italian meal

Families gather together to eat their main meal at midday. The first items on the family table are *antipasti* (appetizers)—cold meats and vegetables that might include salami, olives, and artichoke hearts. Next come *primi piatti* (first courses)—pasta, soup, rice, or *polenta* (cornmeal porridge).

Pasta—a mixture of wheat flour, oil, and water—appears in one form or another on every Italian table. It is an ancient food, even enjoyed by the Etruscans thousands of years ago.

Pasta comes in a great variety of shapes. Long, slender varieties of pasta include *capelli d'angelo* (angel hair) and *vermicelli* (little worms). *Ziti* and *rigatoni* are short, round, hollow pastas. Long, flat pastas include *linguine* (little tongues) and *lingue di passero* (sparrow tongues).

Wheat-flour pasta and olive oil are the foundations of southern Italian food. The north is better suited to grazing livestock, so butter is used more often than olive oil. In the north, rice or polenta is a more common dish than pasta. *Risotto alla milanese* (saffron rice), for example, comes from the northern city of Milan.

After the first courses come *secondi piatti* (second or main courses) of meat, poultry, fish, and eggs. Italy is

A gondolier looks on as diners enjoy a meal at one of Venice's many sidewalk cafes. Italians take great pride in the quality of their food, whether cooked at home or in a trattoria—a small, informal family restaurant.

A woman sells fresh spinach at an outdoor market in Rome, above. Many spinach dishes are described as alla florentine (in the Florentine style). Spinach is also used to enrich pasta dough, and it gives the pasta a green color.

A fish market at Venice's Ponte di Rialto displays the day's catch. Fish from the Mediterranean Sea are an important part of Italian cooking.

A pasta maker shapes the dough—a "paste" of flour, oil, and water—for Italy's staple food. Pasta ranges from long, thin spaghetti to short, fat macaroni. Homemade pasta is usually cooked and eaten as soon as it is made.

noted for its cured meats and sausages, such as *prosciutto* (Parma ham) and *mortadella* (a Bolognese sausage of heavily spiced pork). Fresh meats may be served *alla griglia* (grilled or charcoal-broiled) or *arrosto* (roasted).

Italians enjoy fish of all kinds, including such exotic varieties as *pesce spada* (swordfish) and *calamari* (squid). Such famous specialties as *fritto misto di mare* (fried fish and shellfish) and *zuppa di pesce* (fish soup) include several varieties of fish.

Contorni (vegetables and salads) are next on the Italian menu. When tomatoes were brought to Europe from the New World in the 1500's, Italian cooks greeted them with enthusiasm. Other vegetables include *carciofi* (artichokes) and *piselli* (baby peas).

Dolci (sweets) are the final course of an Italian meal. The best-known Italian dessert, *gelato* (creamy ice cream), may be served with whipped cream, or *affogato* ("drowned" in whiskey). Also popular is *granita*—ice crystals served with fruit syrup.

A Sardinian dessert, *sebadas,* consists of cheese-filled, fried ravioli covered with wild honey. Turin is famous for its fine chocolates, especially the hazelnut-flavored *gianduiotti.* Another traditional dessert is *Baci di Dama* (Lady's Kisses)—a blend of almonds, flour, butter, sugar, liqueur, vanilla, and chocolate.

The finishing touch

After their meal, Italians enjoy coffee, and their imaginative ways with that beverage are appreciated throughout the world. *Espresso,* a strongly flavored coffee drink, is popular, as well as *cappucino,* in which steamed, foaming milk is added to espresso. Milder coffee drinks include *caffe lungo* (with added water) and *caffe macchiato* (espresso with milk).

THE ARTS

Italy has made important contributions to the art, architecture, literature, and theater of the world since the early Middle Ages. But nowhere is the genius of the Italian masters so evident as in the paintings and sculpture of the 1400's and 1500's.

During that time, Italy was made up of about 250 individual cities. Some, like Florence, Venice, and Milan, were rich and powerful *city-states* that operated almost independently from the Holy Roman Empire. Governed by powerful families, the city-states became leading centers of art and learning.

Renaissance painters

In the 1300's and 1400's, the thriving cultural life of Italian cities such as Florence, Milan, and Venice developed into the movement known as the Renaissance. At that time, the Medici family ruled Florence and controlled one of the largest banks in Europe. With their financial backing, the city of Florence supported some of the world's greatest artists.

Michelangelo and Raphael were among the great artists helped by the Medici. When Michelangelo was a young man, his work came to the attention of Lorenzo de' Medici, also known as Lorenzo the Magnificent. Lorenzo invited Michelangelo to stay at his palace, where the great artist began to develop the distinctive style that was to mark his work.

Raphael settled in Florence in 1504, where he studied the paintings of Leonardo da Vinci. In 1508, Pope Julius II asked Raphael to work in Rome, and there he created perhaps his greatest work—the series of frescoes that decorate the pope's private quarters in the Vatican.

Leonardo da Vinci, one of the greatest painters of the Italian Renaissance, was born near Florence, but spent much of his early career as a court artist for Lodovico Sforza, the duke of Milan. When the French overthrew Sforza in 1499, Leonardo left Milan and returned to Florence. There, he and Michelangelo were hired by the Florentine government to decorate the walls of a new government building with scenes of the city's military victories.

Botticelli's *Primavera* shows his interest in beautiful mythological subjects, as well as the clear, rhythmic lines, delicate colors, and poetic feeling that distinguished his work. Botticelli lived and worked in Florence during the late 1400's.

The palace of Duke Federico da Montefeltro who ruled Urbino from 1444 to 1482, is a glorious monument to the Italian Renaissance. Its library is filled with illuminated manuscripts, and its galleries display the works of Uccello, Signorelli, and Titian. In addition to the duke's living quarters, the ducal palace also has a stateroom, theater, two chapels, and a secret courtyard garden.

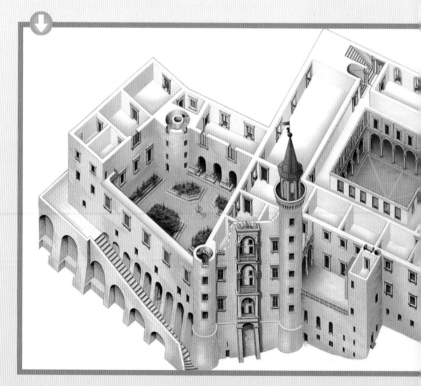

Leonardo da Vinci painted *The Annunciation* in the 1470's for the Convento di S. Bartolomeo at Monteoliveto, just southwest of Florence. The panel now hangs in the Uffizi Gallery in Florence.

Italian architecture

The many beautiful churches throughout Italy stand as proof of the remarkable abilities of the great Italian architects. One example from the Renaissance is the dome for the Cathedral of Florence—a masterpiece of architectural design created by Filippo Brunelleschi.

Perhaps the greatest achievement of Italian Renaissance architecture was St. Peter's Church, originally designed by Donato Bramante in the early 1500's. The building's most outstanding architectural feature is its magnificent dome, designed by Michelangelo.

Literature and music

Italian Renaissance writers produced a number of important works, including *The Prince*—a book that describes the methods a strong ruler might use to gain and keep power—written by Niccolò Machiavelli in 1513.

The Italian love of music may have found its greatest expression in the first operas, which were composed in Florence in the 1590's. The development of opera resulted from the interest of Florentine noblemen, musicians, and poets in the culture of ancient Greece. They believed that the Greeks sang, rather than spoke, their parts in Greek drama. Opera emerged as an art form during the Baroque period of the 1600's and 1700's.

Bernini's chapel altarpiece created for the Cornaro Chapel in Rome, shows an angel poised to drive an arrow into the heart of the swooning Saint Teresa. Bernini's work typifies the highly ornamental and intensely dramatic Baroque style.

The Italian talent for powerful creative expression lives on. Present-day artists, like all the generations that followed the Renaissance and Baroque masters, still use their gifts to create works of extraordinary visual and emotional impact. Inspired by the genius that went before them and surrounded by the masterpieces of the past, modern Italian artists carry on the brilliant tradition that began centuries ago.

AGRICULTURE

Italy has a very long tradition of agriculture. The Etruscans, the ancient people of the Italian Peninsula, introduced sophisticated irrigation techniques into the Po Valley. Later, the Romans became skilled in the growing of fine fruits, especially wine grapes.

In present-day Italy, agriculture no longer plays a major role in the nation's economy. Agriculture, forestry, and fishing now account for only 4 percent of the *gross domestic product* (the total value of goods and services produced within a country in a year), and employ only about 10 percent of Italy's workers.

Most of Italy's farms are individually owned, and about 75 percent of them cover less than 12 acres (5 hectares). Although some of Italy's agriculture has been modernized, much of it remains poor, especially in the south and in the mountain areas.

"Green Plans" for agriculture

After World War II, many Italians left their farms for higher-paying work in the cities. To slow the migration of workers from the land to the factories, the Italian government developed a series of nationwide "Green Plans" in the 1960's and 1970's. The Green Plans attempted to reform the land-tenure system, which had contributed to the development of small, unprofitable farms. The government plans also provided farmers with financial aid for improving agricultural techniques.

The Green Plans helped to increase the land's general productivity. They also encouraged specialization in high-quality produce, such as exotic fruits and vegetables, for both domestic use and export.

Despite the efforts of the government, agriculture has shown little growth in recent decades, and many farms are still small and unprofitable. Although the production of sugar beets, tomatoes, and other crops increased, the

A vineyard worker enjoys a lunch of pasta. The farming population of Italy is aging quickly, and many fear that the younger generation will not take the place of older farmers as they retire.

A herd of sheep graze in the rolling pasturelands of Tuscany. In addition to grazing land, the Tuscan hillsides support much fertile farmland, including large fields of grain, olives, and grapes. The famous Chianti wine is produced in Tuscany.

nation's wheat crop declined. Today, although Italy is a major cereal-producing country, it imports large supplies of wheat, and much of Italy's pasta is made with imported flour.

Grapes, wine, and more

Grapes, Italy's most valuable crop, are grown throughout the country, but the fertile northern valleys produce the best crops. Millions of acres of Italy's fertile land are used for growing grapes.

Italy is the largest producer of wine in the world, and most of its grapes are used for producing wine. Almost every region in Italy produces its own wine, and some wine experts estimate there may be more than 5,000 different Italian wines.

Haymaking provides a store of winter fodder and bedding for livestock in the north of Italy. This area, which includes the lower regions of the Alpine Slope, is protected from extreme cold by the Alps.

A jet of water irrigates a field in Calabria. The southern region of Italy, with its long, hot summers, occasionally suffers drought. The land is used for sheep grazing more than growing crops.

Rice is grown in the well-irrigated Po Valley, Italy's leading agricultural region. The Po Valley is Italy's principal area for livestock and dairy farming. Its crops also include grapes, grains, olives, and sugar beets.

Chianti, perhaps the most familiar Italian wine, comes mainly from Sangiovese grapes native to the regions of Tuscany and Umbria in central Italy. Barolo and Barbaresco, made from Nebbiolo grapes, are wines of the Piedmont region, as is Gavi, a crisp white wine made from Cortese grapes.

Italy also has a large olive crop. Most of the olives are used to make olive oil, another valuable export. Other important crops include rice, barley, corn, oats, and rye. Italy ranks among the world's largest producers of sugar beets, and the country grows a large number of the world's artichokes. It also grows oranges, peaches, apples, tomatoes, and potatoes.

Livestock is raised throughout the country. Northern Italy is noted for its dairy products, beef cattle, and pig and poultry farms. The northern area also provides grazing for large herds of sheep and goats. The Po Valley is famous for the production of fine silk, and mulberry trees are grown there to feed the silkworms.

Coastal and deep-sea fishing in the Mediterranean employs a large number of people and enjoys a bountiful catch. Even so, the industry does not meet Italy's needs because fish is such an important part of the Italian diet.

ECONOMY

After World War II ended in 1945, Italy shifted from an agriculture-based economy to an industry-based economy, and the country entered a period of tremendous economic growth. Today, northern Italy is one of the most advanced industrial areas in Western Europe.

Between 1953 and 1968, the nation's industrial production almost tripled. In 1958, Italy became one of the founding members of the European Economic Community and strengthened its economy through increased trade. The economic boom suffered a setback during a worldwide recession in the 1970's. But by the late 1980's, Italy's economy had recovered enough to place it among the world's leading industrial nations.

Industrial progress in Italy has been achieved despite the country's lack of natural resources. Italy depends heavily on other countries for its energy supply, importing more than half of its petroleum. Large amounts of natural gas from the Po Valley are piped into the cities of the north, and hydroelectric plants in the Alps provide power for northern factories. Hydroelectric plants contribute nearly 20 percent of the nation's electrical supply.

Historically, the government's National Hydrocarbon Agency controlled the production and distribution of Italy's petroleum and gas. The government also owned or controlled much of Italy's business and industry, including steel mills, public utilities, shipbuilding companies, and most of the nation's railroad network. Since the 1990's, however, the government has begun selling many of its holdings to private investors.

Manufacturing

Today, about 25 percent of the Italian work force is employed in industry—a sector that includes manufacturing, mining, construction, and utilities. Clothing, including shoes, is a leading manufactured product in Italy. Many smaller businesses produce textiles and clothing, shoes and leather goods, furniture, and other craft products. Italian silks and wools are prized all over the world.

Colorful leather gloves are displayed in a stylish shop in Venice. Clothing, including accessories, ranks as the leading type of manufactured product in Italy.

New automobiles await shipment from Piedimonte San Germano, south of Rome. Italy traditionally has been a major producer of autumobiles, both passenger cars and high performance sports cars.

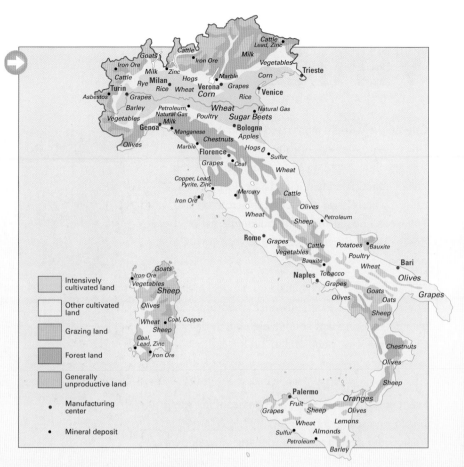

Other important manufactured goods include automobiles, chemicals, electrical and nonelectrical machinery, petroleum products, and processed foods. Most heavy industry is concentrated in northwest Italy, in the triangle formed by Milan, Turin, and Genoa. Major activities in this area include the production of automobiles, household appliances, iron and steel, and machinery and machine tools.

Tourism

Although Italy is limited in natural resources, it is richer than many other countries in natural beauty and historic interest. The artistic and architectural treasures of Rome, Venice, and Florence, the ruins of ancient Pompeii, and the world-class skiing on the Alpine Slopes have all helped make tourism a major industry in Italy.

Millions of tourists visit Italy each year, contributing billions of dollars to the economy. Tourism is a major part of the nation's service industries, keeping people employed in the hotels and restaurants that accommodate visitors.

Looking to the future

Despite Italy's economic achievements, some problems remain. Most postwar industrialization took place in northern Italy, leaving the south lagging behind. Today, southern Italy has a higher unemployment rate than the north and a higher percentage of people who still work in agriculture.

In addition, the worldwide recession at the end of the first decade of the 2000's hit Italy especially hard. By 2011, the nation's huge government debt had become one of the highest in the world. The government was forced to introduce unpopular austerity measures to pay down the debt.

A fashion show in Milan offers a preview of clothing designs. The Italian fashion industry helps set clothing styles throughout the world. Clothing is one of Italy's main exports.

Marble workers shape and polish stone from the famous marble quarries of Carrara, in northern Italy. Renaissance sculptors prized Carrara marble for its white color and compact grain. The stone is cut from the Alpine slopes.

VENICE

One of the world's most famous and unusual cities, Venice lies on a group of islands in the Adriatic Sea, 2-1/2 miles (4 kilometers) off the Italian coast. The main islands are linked to each other by more than 400 bridges across canals. The city's houses and other buildings rest on great posts driven into the mud, rather than on solid ground.

The streets of Venice have an unusual silence about them, due to the absence of automobile traffic. The dominant sounds are church bells and the footsteps of pedestrians.

"La Serenissima"

From earliest times, the Venetian economy was based on fishing and trading. By the A.D. 800's, Venice had developed into a nearly independent city-state. It traded goods with Constantinople (now Istanbul), as well as with cities on the Italian mainland and the northern coast of Africa.

In 1380, Venice defeated Genoa—its rival sea power—and gained control over trade in the eastern Mediterranean Sea. Soon, Venice became one of the world's largest cities, reaching the height of its power during the 1400's.

The city, which called itself "La Serenissima" (the Most Serene Republic), then included Crete, Cyprus, the Dalmatian coast (now part of Croatia), and part of northeastern Italy. Venetian ships carried almost all the silks, spices, and other luxury items that reached Europe from Asia.

Like the leading citizens of other Italian city-states during the Renaissance, the merchants and aristocrats of Venice used some of their great wealth to support the arts. As a result of their patronage, Venice's paintings, textiles, and handicrafts came to be known and prized throughout the civilized world.

In time, the paintings of such Venetian masters as Gentile da Fabriano and Giovanni Bellini, Vittore Carpaccio, and Giorgione took their place among the greatest works in the Western world.

High tides regularly flood Venice's walkways and alleys. Rising waters and sinking land have put the city in great danger, but government funds are now being used for projects to prevent further damage.

Meanwhile, the architects of Venice constructed magnificent palaces and churches along the city's canals.

Venice's golden age came to a halt after the Portuguese explorer Vasco da Gama discovered a sea route to India and Christopher Columbus discovered America in the late 1400's. The center of trade in Europe then shifted to the Atlantic Ocean and the New World. As a result, Venice—once so rich and powerful—declined in importance.

Present-day Venice

Many of the art treasures created during Venice's centuries of wealth and power can still be enjoyed by visitors today. The Basilica of Saint Mark, one of the world's outstanding examples of Byzantine architecture, stands in the southern section of the city. Its interior is richly decorated with mosaics, carvings, and colored marble.

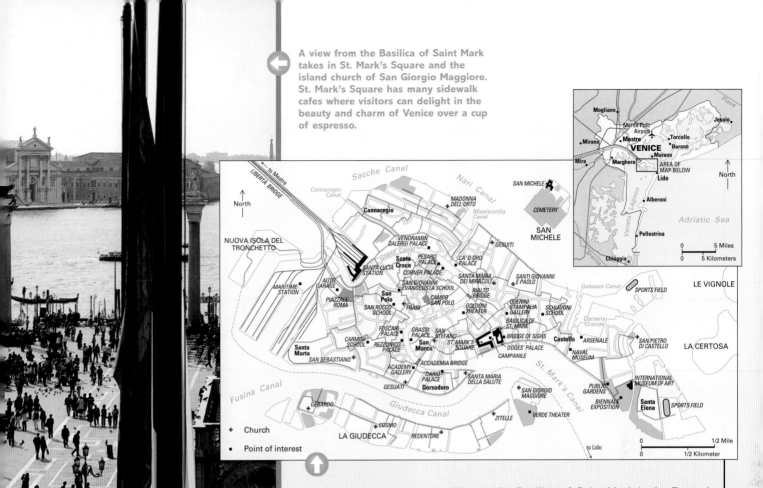

A view from the Basilica of Saint Mark takes in St. Mark's Square and the island church of San Giorgio Maggiore. St. Mark's Square has many sidewalk cafes where visitors can delight in the beauty and charm of Venice over a cup of espresso.

Map labels:

to Mestre · LIBERTA BRIDGE · North · Sacche Canal · Nari Canal · Cannaregio Canal · SAN MICHELE · Miraflora Canal · CEMETERY · SAN MICHELE · MADONNA DELL'ORTO · GESUITI · Cannaregio · NUOVA ISOLA DEL TRONCHETTO · VENDRAMIN CALERGI PALACE · PESARO PALACE · CA' D'ORO PALACE · Grand Canal · MARITIME STATION · AUTO GARAGE · SANTA LUCIA STATION · Santa Croce · CORNER PALACE · SAN GIOVANNI EVANGELISTA SCHOOL · SANTA MARIA DEI MIRACOLI · SANTI GIOVANNI E PAOLO · Galeazze Canal · SPORTS FIELD · LE VIGNOLE · PIAZZALE ROMA · SAN ROCCO SCHOOL · San Polo · Frari · CAMPO SAN POLO · RIALTO BRIDGE · QUERINI STAMPALIA GALLERY · GOLDONI THEATER · SCHIAVONI SCHOOL · Darsena Grande · FOSCARI PALACE · GRASSI PALACE · SAN STEFANO · BASILICA OF ST. MARK · BRIDGE OF SIGHS · ARSENALE · SAN PIETRO DI CASTELLO · LA CERTOSA · CARMINI SCHOOL · REZZONICO PALACE · San Marco · ST. MARK'S SQUARE · DOGES' PALACE · Castello · NAVAL MUSEUM · Santa Marta · SAN SEBASTIANO · CAMPANILE · ACADEMY GALLERY · ACCADEMIA BRIDGE · St. Mark's Canal · PUBLIC GARDENS · INTERNATIONAL MUSEUM OF ART · GESUATI · DARIO PALACE · Dorsoduro · SANTA MARIA DELLA SALUTE · SAN GIORGIO MAGGIORE · BIENNALE EXPOSITION · Santa Elena · SPORTS FIELD · GERARDO · Fusina Canal · Giudecca Canal · COSMO · ZITELLE · VERDE THEATER · to Lido

+ Church
■ Point of interest

LA GIUDECCA · REDENTORE

0 1/2 Mile · 0 1/2 Kilometer

Inset map labels:

Mogliano · Marco Polo Airport · Mestre · Torcello · Jesolo · Mirano · VENICE · Burano · Mira · Marghera · Murano · AREA OF MAP BELOW · Veneta Lagoon · Piave · Adriatic Sea · Lido · Alberoni · Pellestrina · Brenta · Chioggia · North · 0 5 Miles · 0 5 Kilometers

The city of Venice is built on about 120 islands off the north-eastern coast of Italy. About 150 canals connect the islands.

The Rialto Bridge crosses the Grand Canal in the heart of Venice. The present marble causeway, built in the late 1500's, replaced a wooden bridge. Its humped shape allowed the armed Venetian ships to pass underneath.

Next to the Basilica of Saint Mark is the Doges' Palace, a huge pink-and-white Gothic building that was once the home of the rulers of Venice. The palace is joined to the state prison by a narrow, covered bridge called the *Ponte dei Sospiri* (Bridge of Sighs)—named for the unhappy prisoners who crossed it on their way to trial.

A drowning city

The water that gives Venice so much of its charm also threatens to be its downfall. Floods caused by high tides during winter storms, as well as polluted air and water, are now endangering the city.

In 1966, damage due to severe flooding cost millions of dollars and destroyed many of Venice's paintings and statues. The city was sinking by about 1/5 inch (5 millimeters) a year until the 1970's, when laws were passed that restricted the use of water from the city's underground wells. Water pressure then built up under the islands, and the city stopped sinking. In the late 1980's, the government approved billions of dollars for projects to protect the city from further flooding and erosion.

SICILY AND SARDINIA

In addition to the mainland peninsula, the nation of Italy also includes Sicily, the largest island in the Mediterranean Sea, and Sardinia, the second largest island.

Sicily lies off the southwest coast of Italy across the Strait of Messina. Palermo, the capital of Sicily, is the island's center of industry and trade, as well as its chief seaport. A ferry links Palermo with Tunis on the coast of North Africa.

Sardinia lies off the west coast of the Italian mainland, 9 miles (14 kilometers) south of the French island of Corsica in the Tyrhennian Sea. The island of Sardinia and several small islands nearby form the region of Sardinia.

Sicily

More than 85 percent of Sicily is covered by hills and mountains that rise to their highest point of 10,902 feet (3,323 meters) on the snow-capped peak of Mount Etna. Located on the east coast of the island, Mount Etna is one of the world's most famous active volcanoes.

Despite its violent nature, Mount Etna is very beautiful, with its high, snow-capped peak and lush, tree lined slopes. Colorful orchards, vineyards, and orange groves nestle around its base to complete the picture.

Although Mount Etna erupts periodically, the area surrounding the volcano is the most heavily populated region of Sicily. The volcanic ash makes the soil rich and fertile, creating excellent conditions for growing olives, grapes, citrus fruits, and cereals. At night, one can see the red glow of the volcano's lava reflected in the clouds above.

At its narrowest point, the distance between Sicily and mainland Italy is only about 2 miles (3 kilometers)—a short trip on a ferryboat. Yet the traveler who ferries to Sicily from Italy finds a completely different society.

To begin with, Sicilian dialects show traces of Arabic, as well as Greek and other European lan-

The harbor at Alghero in northern Sardinia is crowded with tourists taking in the sights. This charming seaside town shows the Spanish influence in the architecture of its cathedral. Many people who live in Alghero speak a Catalan dialect.

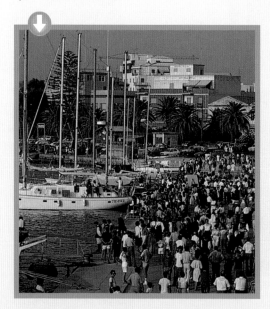

guages. These are among the marks left by centuries of foreign invasion and rule—by Greek colonists as early as the 700's B.C., and later by the Carthaginians, Romans, Germanic tribes, Byzantines, Muslims, Normans, French, Germans, Spanish, and Austrians. Sicily's ancient monuments range from the Greek temples of Zeus and Hera at Agrigento to the Norman Cathedral in Palermo.

Years of foreign rule have left their influence on the Sicilian character as well as on the languages and landscape. Some Sicilians still distrust all forms of government and tend to value personal and family honor over loyalty to their country. This code of honor is known as *omertà* (law of silence), and it forbids telling the police about crimes considered to be private affairs.

Today, Sicily is an island of poor farmers who barely manage to make a living off the land. Despite the Italian government's efforts

to modernize farming, most Sicilian farmers still use old-fashioned methods and equipment. In addition, many people leave Sicily for higher-paying jobs in northern Italy and in other countries.

Sardinia

About 90 percent of Sardinia is mountainous. Very few people live in the mountains, because the steep slopes and heavy rainfall produce landslides and floods.

The only important lowland region is the Campidano plain of the southwest, where almonds, grapes, herbs, lemons, olives, oranges, and wheat are grown. Elsewhere, large herds of sheep and goats are grazed. Although many forests have been cleared, the northeast part of the island still has many cork oak trees, and cork is a leading export. Mines produce copper, iron, lead, lignite, manganese, silver, and zinc.

Industrial development on the island remains limited, but tourism has become an important part of the economy. In recent years, the beautiful Costa Smeralda (Emerald Coast) has become very popular due to the area's beautiful beaches and plentiful tourist accommodations.

Like Sicily, Sardinia has a long history of foreign invasion. Its capital and largest city, Cagliari, was founded by the Phoenicians. Later, the island was settled by the Romans. Sardinia was invaded by Germanic tribes in the 400's, occupied by the Byzantines in the 500's, and attacked by Arabs from the 700's to the 1000's. In 1297, Sardinia was given to the Spanish kingdom of Aragon, but Aragon did not complete its conquest of the island until the end of the 1400's. In 1713, under the Treaty of Utrecht, Austria gained control of the island.

In 1720, along with the Piedmont region on the northwest coast of Italy, the island became a part of the Kingdom of Sardinia. The center of the kingdom's government was in Turin, where the first movement for a united Italy began to take shape in the 1820's and 1830's.

The coastline of Sardinia has a rugged beauty all its own. But, except for its mineral deposits, the island has few natural resources. The island's economy relies on the tourist trade.

The view from the top of the ancient Greek theater in Taormina is one of Sicily's most beautiful. The theater was rebuilt in the Roman style in the A.D. 100's.

JAMAICA

Jamaica, an island nation in the West Indies, is the third largest island in the Caribbean Sea. It lies about 480 miles (772 kilometers) south of Florida, in the tropics. More than 90 percent of Jamaica's 2,758,000 million people have African ancestry. Some of these Jamaicans are *Afro-Europeans*—people of both African and European ancestry. The country's minority groups include Chinese, East Indians and other Asians, Europeans, and Syrians.

Modern life

Most Jamaican professional and business people are Europeans or Afro-Europeans. Many Chinese and Syrians are shopkeepers. Large numbers of blacks and Asians work as farm laborers.

Nearly half of Jamaica's people live in rural areas, but many have moved to the cities since the early 1960's. Jamaica's official language is English, but the people actually speak a *dialect* (local form) of English that differs from the language spoken by American or British people.

About two-thirds of Jamaicans are Christians. About 100,000 Jamaicans belong to a religious and political movement called Rastafarianism. Rastafarians, as members are called, consider former Emperor Haile Selassie I of Ethiopia a god. Rastafarians have also adopted many of the beliefs of Marcus Garvey, a Jamaican who died in 1940. Garvey preached that all blacks should consider Africa their home and live there. Rastafarians consider themselves Africans, not Jamaicans. Some believe blacks are a superior race.

History

Arawak Indians were living in what is now Jamaica when Christopher Columbus landed there in 1494. Columbus claimed the land for Spain. The Spaniards enslaved the Arawak people, and almost all the Indians were killed by diseases or overwork. The Spaniards then brought Africans to Jamaica to work as slaves. Spain did not try to settle or develop the island. Instead, it used Jamaica mainly as a supply base.

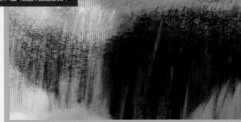

Balancing bananas on her head, a Jamaican woman helps bring in the harvest. About one-fourth of Jamaica's workers, especially those of African and Asian ancestry, are farmworkers.

Cricket fans are guaranteed an unobstructed view of the game from high atop fence posts. Cleverly placed coconuts add comfort to their enjoyment of Jamaica's national sport.

Waterfall climbing is a sport unique to Jamaica. Island of Springs was the original Arawak Indian name for the island.

The British invaded Jamaica in 1655 and controlled most of the island within five years. However, they had to keep fighting escaped African slaves called *Maroons,* who had fled to the rugged hill region of Jamaica, called the Cockpit Country, when the British arrived. During the 1670's, British pirates used Jamaica as a base to attack Spanish ports and ships in the Caribbean.

Jamaica prospered during the 1700's. Sugar cane became the major crop. The island also ranked as the most important slave market in the Western Hemisphere. In 1738, the British finally signed a peace treaty with the Maroons, on the former slaves' terms. About a hundred years later, the British Parliament freed all the slaves on the island. Most of the freed slaves became independent farmers. However, the sugar industry was hurt because the plantation owners lost thousands of slave laborers.

In 1865, relations between the planters and their workers had deteriorated so much that the workers revolted. The Morant Bay Rebellion, as the revolt was called, was led by Paul Bogle, a Baptist deacon. British troops put down the uprising. The British government then made Jamaica a crown colony, governed entirely by the United Kingdom. They eliminated the House of Assembly that Jamaicans previously had elected to help the British rule the island.

During the 1930's, Jamaican labor leaders urged the United Kingdom to grant Jamaicans more political power. In 1944, a new constitution gave Jamaicans some self-government. In 1962, Jamaica achieved full independence.

A cluster of shacks on the outskirts of Montego Bay reflects the continuing movement of poor rural people to the towns.

JAMAICA TODAY

The Arawak Indians, who were the first people to live in Jamaica, called the island *Xaymaca,* which meant *land of wood and water.* Today, Jamaica's beautiful beaches, soaring mountains, and pleasant climate attract large numbers of tourists every year.

The land and climate

Jamaica's many mountains include the Mocho Mountains, which rise in the center of the island, and the Blue Mountains, which rise in the east. Swift rivers flow north and south from the mountains, and the island also has many springs, streams, and waterfalls.

Limestone formations are found in the northwest area known as Cockpit Country. This region was named for the many deep depressions called *cockpits.* Low plains along the coasts are lined with many beautiful beaches. Jamaica has a tropical climate, but the heat and humidity are moderated by ocean breezes along the coasts and by altitude in the mountains. Temperatures sometimes drop to 40° F (4° C) in mountain areas.

The economy

Thanks to Jamaica's tropical beauty and pleasant climate, tourism is one of the country's leading economic activities. Tourist centers include Kingston, Montego Bay, Negril, and Ocho Rios.

Jamaica is among the world's leading producers of bauxite, and mining provides much of the nation's income. Plants near some of the bauxite mines remove a mineral compound called *alumina* from the bauxite ore—the first step in the production of aluminum. Jamaicans also mine gypsum, which is used in making plasterboard and other construction materials. Minerals are the chief exports.

Although agriculture employs about a fourth of all Jamaican workers, the farms do not produce enough food for the country, so Jamaica imports

FACTS

Official name:	Jamaica
Capital:	Kingston
Terrain:	Mostly mountains, with narrow, discontinuous coastal plain
Area:	4,244 mi² (10,991 km²)
Climate:	Tropical; hot, humid; temperate interior
Main rivers:	Great, Black, Minho, Cobre, Yallahs
Highest elevation:	Blue Mountain Peak, 7,402 ft (2,256 m)
Lowest elevation:	Caribbean Sea, sea level
Form of government:	Constitutional monarchy
Head of state:	British monarch, represented by governor general
Head of government:	Prime minister
Administrative areas:	12 parishes and 1 governmental unit (Kingston and its surrounding area)
Legislature:	Parliament consisting of the Senate with 21 members and the House of Representatives with 60 members serving five-year terms
Court system:	Supreme Court, Court of Appeal
Armed forces:	2,800 troops
National holiday:	Independence Day - August 6 (1962)
Estimated 2010 population:	2,758,000
Population density:	650 persons per mi² (251 per km²)
Population distribution:	53% urban, 47% rural
Life expectancy in years:	Male, 71; female, 75
Doctors per 1,000 people:	0.9
Birth rate per 1,000:	17
Death rate per 1,000:	6
Infant mortality:	21 deaths per 1,000 live births
Age structure:	0-14: 31%; 15-64: 61%; 65 and over: 8%
Internet users per 100 people:	57
Internet code:	.jm
Languages spoken:	English, English patois
Religions:	Protestant 62.4%, Roman Catholic 2.6%, other 35%
Currency:	Jamaican dollar
Gross domestic product (GDP) in 2008:	$14.31 billion U.S.
Real annual growth rate (2008):	-0.6%
GDP per capita (2008):	$5,310 U.S.
Goods exported:	Alumina, coffee, petroleum products, rum, sugar
Goods imported:	Chemicals, crude oil and petroleum products, food, machinery, motor vehicles
Trading partners:	Canada, United Kingdom, United States, Venezuela

much of its food supply. Sugar cane is the nation's most important crop. Other farm products include allspice, bananas, cacao, citrus fruits, coconuts, coffee, milk, poultry, and yams. Jamaica manufactures cement, chemicals, cigars, clothing, fertilizer, footwear, machinery, molasses, petroleum products, rum, and tires.

The government

Jamaica was a British colony for about 300 years. Today, the country is an independent nation within the Commonwealth of Nations. The head of state is the monarch of the United Kingdom, represented by a governor general. The governor general has little power, however. Jamaica's chief executive is its prime minister—the leader of the majority party in Parliament, Jamaica's legislature.

The Jamaican Parliament consists of a House of Representatives and a Senate. The 60 representatives are elected by the people to five-year terms. The 21 senators are appointed by the governor general, 13 of them on the advice of the prime minister and 8 on the advice of the leader of the minority, or opposition, party in Parliament. The largest political parties are the Jamaican Labor Party (JLP) and the People's National Party (PNP).

Jamaica became independent in 1962. Since then, the country has faced many problems, including poverty, unemployment, and inflation. In the 1970's, Michael Manley of the PNP became prime minister. He tried to solve the economic problems by adopting socialistic policies. He also began seeking relations with leftist governments, a policy that worried Jamaica's traditional Western allies, such as the United Kingdom and the United States. Jamaica's tourist industry suffered.

Edward Seaga of the JLP became prime minister in 1980 and adopted policies designed to help private business and promote good relations with Western nations. In 1989, Manley became prime minister again, but stated he would follow more moderate economic and foreign policies. Manley resigned in 1992. Percival J. Patterson was elected head of the PNP and became prime minister.

In 2006, the PNP's Portia Simpson Miller became Jamaica's first woman prime minister. In general elections in 2007, the PNP was defeated by the JLP. Bruce Golding became prime minister and tried to deal with Jamaica's continuing problems of economic stagnation and violent crime. He stepped down in 2011. The PNP won the general election, and Miller again became prime minister.

The island of Jamaica is the third largest Caribbean island. Only Cuba and Hispaniola are larger. Kingston is Jamaica's capital, largest city, and chief port.

Red-stained water, caused by the mining of bauxite, mars the natural beauty of Jamaica. Jamaica is among the world's leading producers of bauxite, the ore from which aluminum is made.

REGGAE MUSIC

Jamaica has long been known for its tropical beauty and pleasing climate. It has also become known as the birthplace of a popular style of music.

Many of the Caribbean islands have their own special kind of music. In Cuba, the music is salsa, and in Trinidad, it is calypso. In Jamaica, the style of music is reggae. For many Jamaicans, however, reggae is more than just a local music; it is closely connected with politics, religion, and pride in their African roots.

Reggae arose in Jamaica from many different kinds of music. It grew out of Jamaican folk music as well as ska and rock-steady, which, in turn, grew out of American rhythm and blues and soul. Reggae's regular, steady rhythm can be traced to African music. The basic instrumental arrangement of electric bass, rhythm guitar, keyboard, drum kit, and horns is accompanied by a distinctive kind of singing.

Reggae began to develop in the 1960's, as young musicians in Kingston sought to reflect the tensions of the times in their music. Jamaica had achieved independence, but poverty and unemployment were widespread. Many reggae songs demanded social and political changes. Local Jamaican bands made the music popular on the island, and politicians came to use reggae music and musicians in their struggle for supporters.

Reggae gained international popularity largely through a musician and singer named Bob Marley. Born in rural Jamaica in 1945, Marley began singing and writing music in his teens. His group, the Wailers, became one of the most popular reggae bands in Jamaica in the early 1960's and had become famous outside Jamaica by 1973. Many of his songs express the beliefs of Rastafarianism, a religious and political movement to which Marley belonged. Marley died of cancer in 1981, but he had already become a legendary figure both in Jamaica and elsewhere.

Admirers are welcomed at both Marley's recording studio and his former home, which opened as a museum dedicated to him in 1986. A mural depicting his life is included in a tour of the museum. A heart in the center of the mural represents the heart of the Jamaican people and the love they have for Marley.

Ziggy Marley & the Melody Makers is led by Bob Marley's son.

Kingston's Trench Town is the birthplace of many reggae bands.

The Reggae Sunsplash festival was held every year in Montego Bay. It featured a variety of reggae artists, including Lieutenant Stitchie. Sunsplash was once the best-known reggae festival.

Rastafarians also believe in the teachings of Marcus Garvey, a black leader who started a "back to Africa" movement. Garvey was born in Jamaica and started his movement there, then moved to the United States. Garvey believed that blacks would never receive justice in countries where most of the people were white. He preached that blacks should consider Africa their homeland.

Reggae music arose from the ethnic pride of the Rastafarian movement. The reggae music of Bob Marley overcame racial and cultural barriers to become popular around the world.

With his popularity, Marley influenced many musicians. For instance, Eric Clapton's version of "I Shot the Sheriff" became a hit in the United States. In addition, Marley's children carried on his music. His son David "Ziggy" Marley achieved success with his band Ziggy Marley & the Melody Makers. Marley's children have also performed in other groups, including the Ghetto Youths Crew and the Marley Girls.

The reggae sound is heard in nightspots throughout Jamaica. The Jamaican tourist spot of Montego Bay held a concert festival called Reggae Sunsplash for many years. Currently, Jamaica has a reggae festival every summer called Reggae Sunfest. Reggae music festivals are also popular in many other countries.

Jamaica was the birthplace of reggae. Other West Indian islanders have produced their own kinds of music.

Located in Kingston, the Bob Marley Museum is easily recognized because red, yellow, and green stripes mark the entrance, and the Ethiopian flag flies outside. Red, yellow, and green are the colors of Rastafarianism.

The Rastafarian movement began in Jamaica in the 1920's. It was named for Ras (Duke) Tafari Makonnen, who took the title Haile Selassie I when he became emperor of Ethiopia in 1930. Several Jamaican religious sects had predicted that a black emperor would be a savior of black people around the world. Rastafarians consider Selassie to be a god. Rastafarian practices include avoiding foods considered impure and not cutting the hair—leading to the ropelike braided hair style called *dreadlocks*.

Robert Nesta Marley, known as Bob Marley, made reggae music popular around the world. The music became a vehicle for black pride. In recognition of his achievement, Jamaica honored him with the Order of Merit award in 1981.

Japan, a beautiful island country in the North Pacific Ocean, lies off the eastern coast of mainland Asia and faces southeastern Russia, the Korean Peninsula, and China. Japan's four major islands—Hokkaido, Honshu, Kyushu, and Shikoku—and thousands of smaller ones form a curved landmass that extends for about 1,200 miles (1,900 kilometers). More than 127 million people are crowded onto these islands, making Japan—with 875 people per square mile (338 per square kilometer)—one of the world's most densely populated countries.

The Japanese call their country Nippon or Nihon, which means *source of the sun*. The name *Japan* may derive from *Zipangu,* a name given to the country by the Italian trader Marco Polo. Polo heard about the Japanese islands while traveling throughout China in the 1200's.

Framed by cherry blossoms, majestic Mount Fuji symbolizes the great natural beauty of Japan. Because mountains and hills take up most of the land, the great majority of the Japanese people live on the narrow coastal plains. These plains have much of the country's best farmland and most of its largest cities, which are centers of commerce, culture, and industry. Tokyo, the country's largest city, is also its capital. Most of Japan's people live in urban areas.

Early Japan was greatly influenced by China and borrowed heavily from Chinese art, government, language, religion, and technology. Japan closed its doors to Western influence in the early 1600's but renewed relations with the West in the mid-1800's. By the early 1900's, Japan had become a leading industrial and military power. However, the country suffered a devastating defeat in World War II (1939-1945). By 1945, many Japanese cities lay in ruin, industries were shattered, and Allied forces occupied the country.

The Japanese worked hard to rebuild their country, concentrating their efforts on economic development. By the late 1960's, Japan had become a great industrial nation. Today, the country is one of the world's economic giants with a total economic output exceeded only by that of the United States and China. Japan manufactures such products as automobiles, electronic goods, and communications and data processing equipment. Its factories have some of the most advanced equipment in the world. However, Japan has few natural resources and must import many of the raw materials needed for its industry.

Japan's big cities have many similarities to those of Western nations, and the Japanese people enjoy a high standard of living. Life in Japan reflects the culture of both the East and the West. For example, baseball games and exhibitions of traditional sumo wrestling are the nation's favorite sporting events.

JAPAN TODAY

World War II (1939-1945) left Japan completely defeated, but the Japanese people worked hard to bring about the country's recovery. By the 1970's, Japan had become a great industrial nation. The success of the Japanese economy attracted attention throughout the world. Today, few nations enjoy a standard of living as high as Japan's.

Japan's government underwent major changes after World War II. In 1947, a new constitution, drawn up by the Allied occupation forces, transferred all political power from the Japanese emperor to the Japanese people. The Constitution also took away Japan's right to wage war and abolished its armed forces. However, Japan now maintains air, ground, and sea forces for purposes of self-defense.

The Constitution guarantees the Japanese people many rights, including freedom of religion, speech, and the press. It also provides for three branches of government—legislative, executive, and judicial. The legislative branch is composed of a two-house parliament called the *Diet*. The Diet makes Japan's laws. In addition, its members choose the prime minister, the country's chief executive. The prime minister selects members of the Cabinet to help govern the country. The Supreme Court, Japan's highest court, consists of a chief justice and 14 associate justices.

Japan has several political parties. The most important include the Liberal Democratic Party (LDP), the Democratic Party of Japan (DPJ), the Japan Communist Party, the Social Democratic Party, and New Komeito (also called the New Clean Government Party).

Japan's Emperor Hirohito, who had reigned since 1926, died in January 1989. His son, Akihito, assumed the throne upon the emperor's death and pledged to protect the Japanese Constitution.

Japan faced economic troubles in the 1980's and 1990's, as problems in trade and finance resulted in a recession. The country's unemployment rate rose, average household incomes nearly stopped growing, and consumer spending declined. However, other nations also suffered economic problems, so Japan's economic position in the world did not change dramatically. Economic challenges continued in the early 2000's.

In 1993, the Liberal Democratic Party, which had ruled Japan since the party was founded in 1955, lost its majority in parliament for the first time. The LDP returned to power in 1994. It controlled the government in the early 2000's until the Democratic Party of Japan defeated the LDP in parliamentary elections in 2009.

FACTS

Official name:	Japan
Capital:	Tokyo
Terrain:	Mostly rugged and mountainous
Area:	145,914 mi² (377,915 km²)
Climate:	Varies from tropical in south to cool temperate in north
Main rivers:	Ishikari, Shinano, Tone
Highest elevation:	Mount Fuji, 12,388 ft (3,776 m)
Lowest elevation:	Sea level
Form of government:	Parliamentary democracy with ceremonial emperor
Head of state:	Emperor
Head of government:	Prime minister
Administrative areas:	47 prefectures
Legislature:	Kokkai (Diet) consisting of the Sangi-in (House of Councillors) with 242 members serving six-year terms and Shugi-in (House of Representatives) with 480 members serving four-year terms
Court system:	Supreme Court
Armed forces:	230,300 troops
National holiday:	Birthday of Emperor Akihito - December 23 (1933)
Estimated 2010 population:	127,669,000
Population density:	875 persons per mi² (338 per km²)
Population distribution:	67% urban, 33% rural
Life expectancy in years:	Male, 79; female, 86
Doctors per 1,000 people:	2.1
Birth rate per 1,000:	9
Death rate per 1,000:	9
Infant mortality:	3 deaths per 1,000 live births
Age structure:	0-14: 14%; 15-64: 64%; 65 and over: 22%
Internet users per 100 people:	69
Internet code:	.jp
Language spoken:	Japanese
Religions:	Shinto, Buddhist, Christian (Note: many people practice both Shintoism and Buddhism)
Currency:	Yen
Gross domestic product (GDP) in 2008:	$4.892 trillion U.S.
Real annual growth rate (2008):	-0.4%
GDP per capita (2008):	$38,223 U.S.
Goods exported:	Chemicals, electronic equipment, machinery, motor vehicles, scientific and optical equipment, semiconductors
Goods imported:	Chemicals, crude oil, fish and shellfish, machinery, natural gas, transportation equipment
Trading partners:	Australia, China, Germany, South Korea, Taiwan, United Arab Emirates, United States

On March 11, 2011, Japan experienced one of the worst disasters in its history. A powerful earthquake occurred off the coast of Honshu, Japan's largest island. The earthquake caused a *tsunami* (series of powerful ocean waves) that struck Honshu's northeastern coast. Waves as high as 33 feet (10 meters) washed away tens of thousands of buildings and vehicles and flooded acres of farmland. More than 15,800 people were killed, and more than 3,400 others were missing.

The earthquake and tsunami damaged a nuclear power plant on Honshu. Water from the tsunami damaged cooling systems at the plant, causing a meltdown of nuclear fuel rods in multiple reactors. Low-level radiation spread throughout the countryside for several weeks after the disaster.

The government's response to the crisis was heavily criticized. Naoto Kan of the DPJ, who had become prime minister in June 2010, stepped down in August 2011. The DPJ chose Finance Minister Yoshihiko Noda to replace him.

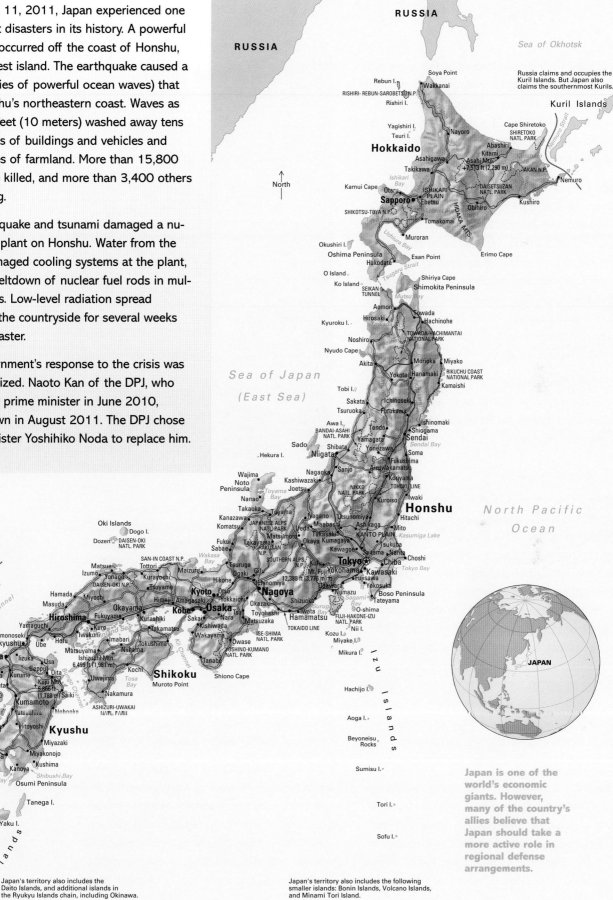

Japan is one of the world's economic giants. However, many of the country's allies believe that Japan should take a more active role in regional defense arrangements.

Japan's territory also includes the Daito Islands, and additional islands in the Ryukyu Islands chain, including Okinawa.

Japan's territory also includes the following smaller islands: Bonin Islands, Volcano Islands, and Minami Tori Island.

ENVIRONMENT

Mount Aso, a volcano on the island of Kyushu, is surrounded by lush green fields and scattered farming settlements. Lakes have formed in the craters of some extinct volcanoes.

The Japanese islands are formed by the upper part of a great mountain range that rises from the floor of the North Pacific Ocean. Mountains and hills cover about 70 percent of Japan's land area, and dense forests cover much of the mountainsides, adding to the beauty of the Japanese islands.

Japan lies on an extremely unstable part of Earth's crust and is therefore subject to frequent earthquakes and volcanic activity. Most of the roughly 1,500 earthquakes that Japan experiences every year are minor, but severe quakes occur every few years. In addition, undersea earthquakes sometimes cause huge, destructive ocean waves, called *tsunamis,* along Japan's Pacific coast. The Japanese islands also have more than 150 major volcanoes, over 60 of which are active.

Japan is composed of four main islands and thousands of smaller islands and islets. In order of size, the main islands are Honshu, Hokkaido, Kyushu, and Shikoku.

The main islands

Honshu, Japan's largest island, is home to about 80 percent of the Japanese people. Three mountain ranges cross northern Honshu, and two plains that support farming lie to the east and west of these mountains. The Japanese Alps, Japan's highest mountains, rise in central Honshu. A chain of volcanoes that cuts across the center of the island includes Mount Fuji, an inactive volcano and the country's tallest and most famous peak. Farther east lies the Kanto Plain, the country's largest lowland area as well as an important center of agriculture and industry. Tokyo stands on the Kanto Plain. Two other major agricultural and industrial lowlands lie south and west of the Kanto region. Mountains cover most of southwestern Honshu.

Hokkaido is the northernmost of Japan's four major islands. It is the country's second largest island in area but has only about 4 percent of the total population. The Ishikari Plain in west-central Hokkaido is the largest lowland and the chief farming region on the island. Smaller plains border the

island's east coast, and much of the rest of Hokkaido consists of forested mountains and hills. The island is a popular area for winter sports.

Kyushu, the southernmost of the main islands, has the second highest population after Honshu, with about 10 percent of Japan's population. Most of the people live in the heavily industrialized lowlands of northwestern Kyushu and in the major farming districts along the west coast. A chain of steep, heavily wooded mountains runs down the center of the island. The volcanic regions in the northeastern and southern sections of Kyushu contain only small patches of farmland.

Shikoku, the smallest of the main Japanese islands, lies off the coast of southwestern Honshu and has only about 3 percent of Japan's total population. Most of the people on this largely mountainous island live in its northern section, where the land slopes down to the Inland Sea. Farmers grow rice and a variety of fruits on the

fertile land along the Inland Sea, and some copper mining is carried out. Hundreds of hilly, wooded islands dot this body of water. A narrow border along Shikoku's southern coast also supports some farming.

Climate

Seasonal monsoons and two Pacific Ocean currents—the Japan Current and the Oyashio Current—affect Japan's climate. Hokkaido and northern Honshu, influenced by the cold Oyashio Current and by monsoons from the northwest in winter, have cool summers and severe, snowy winters. The southern and eastern areas of the country are warmed by the Japan Current and by monsoons from the southeast in summer. These areas have mild winters and warm, humid summers.

All areas of the country—except eastern Hokkaido—receive at least 40 inches (100 centimeters) of rain yearly, mainly in the summer and early fall. In addition, several typhoons hit Japan each year, chiefly in late summer and early fall. These storms often do great damage to houses and crops.

Tiny offshore islets, as well as large mountainous islands, make up Japan's landscape. Most of Japan's people live near the coasts. The mountainous interiors of the islands are thinly populated.

Nikko National Park on Honshu has a magnificent scenic landscape with high peaks, rocky gorges, and thundering mountain waterfalls. Many vacationers visit the park to enjoy some of Japan's most spectacular scenery.

The hourly progress of a tsunami produced by an earthquake that originated south of Alaska is shown on the map. Tsunami is a Japanese word meaning harbor wave. These waves generally occur along fault lines in the Pacific Ocean. A tsunami can travel 500 to 600 miles (800 to 970 kilometers) per hour in open waters. As the tsunami approaches shallower water near shore, it may form a gigantic wall of water more than 100 feet (30 meters) high.

During an undersea earthquake, the shifting rocks of the sea floor create a wave in the water above. The wave may begin small in deep ocean water, but it can develop into a towering tsunami as it nears the shore.

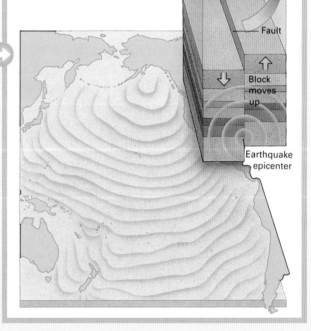

ISLAND GROUPS

The Bonin, Izu, Ryukyu, and Volcano island chains are among the thousands of smaller islands that make up Japan's land area. Although Japan also claims several islands in the southern part of the Kurils, Japanese control of these islands was taken away at the end of World War II. The southern Kuril Islands, which lie off the northeast coast of Hokkaido Island, are referred to by the Japanese as the Northern Territories.

The Izu Islands are a group of volcanic islands that stretch southward off the coast of Honshu. Most of the people on these islands work on farms or fish in coastal waters. Farther south lie the three islands that comprise the sparsely populated Volcano Islands. Iwo Jima (now called Iwo To), the middle island, was captured from the Japanese by American forces during World War II. The United States controlled the island until 1968, when it was returned to Japan.

The Bonin Islands lie between the Izu and Volcano island chains, about 600 miles (970 kilometers) south of Tokyo. The 97 volcanic islands in the Bonin Island chain have a total area of about 41 square miles (106 square kilometers). The rocky, rugged land on the islands is covered with scrubby trees and tall grass. About 2,700 people live on the islands. They raise cacao, cattle, fruits, sugar cane, and vegetables. They also make coral ornaments.

Japan claimed the Bonins in 1875 and named them *Ogasawara-gunto*. American forces attacked the Japanese stationed on the Bonins during World War II. The islands were placed under U.S. control after the war, but they were returned to Japan in 1968. Today, visitors enjoy the islands' warm, mild climate and white sand beaches.

The Ryukyu Islands are a group of more than 100 islands that extend from the southern tip of Kyushu Island to Taiwan. They have a land area of 1,205 square miles (3,120 square kilometers) and a population of about 1-1/2 million. Some of the islands, particularly those containing active volcanoes, are uninhabited.

Most of the Ryukyuans are farmers. They grow rice, but their main food crop is sweet potatoes. They also raise and export sugar cane and pineapples. Fishing is another important activity for the Ryukyuans, bringing food and

Sugar cane is one of the chief crops raised on Okinawa as well as on the other islands of the Ryukyu chain. Although agriculture is important to the economy of the islands, fishing and tourism also produce income.

income. The people speak a language similar to Japanese, and their religious beliefs have been heavily influenced by both Japan and China.

Okinawa, the largest and most important of the Ryukyu Islands, covers an area of 554 square miles (1,434 square kilometers) and has about 1 million people. Naha, the capital and largest city of the islands, is on Okinawa. Like the rest of the islands in the Ryukyu chain, Okinawa is largely mountainous and has a warm, wet climate. Farmers on Okinawa raise much the same crops grown throughout the Ryukyus, but the island's economy depends largely on tourism and United States military spending.

Japan and China both claimed Okinawa and the rest of the Ryukyus until 1874, when China signed a treaty recognizing Japanese rule. One of the bloodiest battles of World War II was fought on Okinawa between U.S. and Japanese troops in 1945. More than 50,000 American troops were killed and about 110,000 Japanese died. About 90 percent of the island's buildings were destroyed during the campaign.

↑

The beautiful beaches and warm weather of the Bonin Islands attracts many tourists. The Japanese government declared the Bonins a national park region in 1972 to protects the islands' natural beauty.

Japan consists of four main islands and thousands of smaller islands. The main islands, in order of size, are Honshu, Hokkaido, Kyushu, and Shikoku. The Ryukyu Islands to the south also belong to Japan. Okinawa is the largest of the Ryukyu Islands.

The modern town hall in Nago on Okinawa was constructed after World War II. Most of the island's buildings were destroyed during the war and had to be replaced.

After Japan was defeated, the United States took over the Ryukyus. In 1953, the United States returned the northern islands to Japan, and Okinawa and the southern Ryukyus were returned in 1972. Under an agreement between the United States and Japan, U.S. military bases remain on Okinawa, but nuclear weapons may not be kept on the island without Japan's consent.

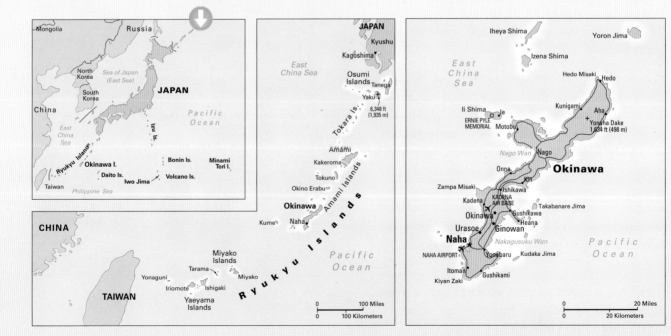

AGRICULTURE

As a result of Japan's massive industrial development after World War II, the percentage of the total labor force employed in agriculture fell from about 45 percent in 1950 to about 4 percent today. Today, agriculture accounts for only 1 percent of Japan's *gross domestic product,* the total value of all goods and services produced within a country in one year.

Although only about 15 percent of Japan's land can be cultivated, Japanese farmers produce much of the food needed to feed the nation's people. Farms average only a few acres, but they produce extremely high yields. Over the years, land reform and modern technology have increased agricultural production and reduced the number of workers required. Japanese farmers make their land as productive as possible through the use of irrigation, improved seed varieties, and modern agricultural chemicals and machinery. In the country's mountainous regions, farmers grow some crops on terraced fields.

Before World War II, many Japanese farmers rented their land and paid their landlords as much as 50 percent of their harvest. After the war, a land-reform program reduced the holdings of the landlords and enabled the farmers to buy the land they worked. Today, most of Japan's farmers own their farms.

More than 50 percent of Japan's farmland is used to grow rice, the country's staple food. Rice is grown throughout the Japanese islands, and production exceeds demand.

Japanese farmers also raise a wide variety of other crops, including sugar beets, tea, tobacco, and wheat. The leaves of mulberry bushes, which are grown on some hillsides, feed silkworms used in the production of raw silk. Apples, cabbages, citrus fruits, eggplants, pears, potatoes, strawberries, and tomatoes are among the fruits and vegetables grown in Japan. In addition, because the Japanese people have increased their consumption of meat and dairy products, more farmers are raising beef and dairy cattle, chickens, and hogs.

Fish has provided the chief source of protein in the Japanese diet for centuries. Japan's fishing industry

Japanese farmers work in rice fields on the southern island of Shikoku. Farmers grow rice and a variety of fruits and vegetables on the fertile land along the island's coasts.

Japanese women pick tea leaves on an emerald-green hillside. Because Japan is so mountainous, level farmland is scarce. As a result, many Japanese farmers grow crops on terraced fields—that is, level strips of land cut out of the hillsides.

employs only about 1 percent of the country's total labor force, yet it is one of the largest in the world. Japan has hundreds of thousands of fishing vessels, and they catch millions of short tons (millions of metric tons) of fish yearly.

Japan ranks among the leading nations in tuna and salmon fishing. Flatfish, mackerel, pollock, and sardines are also important to the nation's fishing industry. In addition, Japanese fishing crews catch large quantities of shellfish and harvest oysters and edible seaweed from "farms" in coastal waters.

Japan's fishing fleets operate all over the world. However, their activities have been limited since the 1970's, when almost all coastal nations established protected fishing zones extending 200 nautical miles (370 kilometers) off their shores. These protected zones excluded Japanese fleets from some of their valuable fishing grounds and thereby reduced Japan's annual catch.

Industrial pollution in Japan's own coastal waters has further reduced the catch. As a result of this decline in production, Japan has had to import some seafood to meet its needs.

For many years, Japan was a leading whaling nation. International concern over endangered whales, however, eventually persuaded Japan to limit its whale catch. Finally, in 1988, Japan agreed to join the other countries of the world in imposing a halt on all commercial whaling.

Dock workers handle a huge tuna catch at a Japanese port. To satisfy the Japanese people's appetite for seafood, Japan operates one of the world's largest fishing industries.

INDUSTRY

Japan began its industrialization process in 1868 under the Meiji leaders, when the government established a number of experimental industrial projects to serve as models for new industries. American and European experts in many fields were hired to teach Western knowledge and methods to the Japanese people. By the early 1900's, Japan had achieved significant industrial growth and developed some foreign trade.

By the 1920's, Japan's most important industries were under private ownership. Financial institutions and manufacturing, mining, and trading companies were controlled by the *zaibatsu,* huge corporations owned by single families. In the late 1930's and early 1940's, the output of the country's modern industries doubled as Japan prepared for war.

Japan's defeat in World War II, however, severely damaged its economy, and it destroyed half of the country's industrial capacity. But the Japanese amazed the world by quickly overcoming the effects of the war. Japan's recovery was aided in large part by financial aid from the United States. The Japanese invested in the latest technology to make their postwar industries highly productive, and labor unions were established. By the late 1960's, Japan was a leading industrial nation.

Manufacturing and service industries

Manufacturing, which employs about 20 percent of the country's workers, is an extremely important economic activity in Japan. It accounts for approximately 20 percent of the country's gross domestic product.

The production of transportation equipment is one of the country's most important manufacturing industries. Japan is among the world's leading producers of ships, cars, and trucks, and it ranks among the top producers of steel. Japan's thriving chemical industry produces petrochemicals and petrochemical products, such as plastics and synthetic fibers. In addition, Japan also produces ceramics, clothing, fabricated metal products, and food products.

One of the fastest-growing industries in the nation is the production of machinery, including heavy electrical

Japan's manufactured products range from tiny computer components to giant oceangoing ships. The country imports many raw materials to meet the needs of its manufacturing industries.

and nonelectrical machines, electrical appliances, and electronic equipment. Japan is also noted for its production of precision instruments, such as cameras, clocks, and watches.

Service industries, which employ about 70 percent of the country's workers, are another important economic activity in Japan. Service industries include banks and other financial institutions, government agencies, trade, transportation, and communication. Altogether, these economic activities account for 70 percent of Japan's gross domestic product.

Robots assemble a sports utility vehicle (SUV) at a modern factory in Kanda Town. Cars and trucks rank among Japan's most important manufactured products.

Video game consoles sit on display in a store in Soka City. Japan is an important international producer of video games as well as such electronics products as computers and television sets.

Foreign trade

Because Japan has limited natural resources, it must import many of the raw materials needed by its manufacturing industries. Japan sells its manufactured goods throughout the world to pay for its imports. Thus, the Japanese economy depends heavily on foreign trade.

Since the mid-1960's, the country has usually maintained a favorable *balance of trade* (the difference between the value of a nation's exports and the value of its imports). However, in the late 1970's and early 1980's, many of Japan's trading partners criticized the country's trade policies. They claimed that while Japan exported large quantities of competitively priced goods to foreign markets, it imposed trade barriers on imports from other countries. These policies were creating unfavorable trade balances for Japan's trading partners, especially the United States.

To maintain good trade relations, Japan began to limit some of its exports and to lift obstacles to imports. Nevertheless, some U.S. officials have continued to criticize the imbalance that remains in Japan's favor.

The bullet-shaped train called the Shinkansen is one of the fastest trains in the world. Japan has a modern, highly efficient transportation system, including airports, highways, railroads, and coastal shipping.

The Tokyo Stock Exchange is the largest stock exchange in Japan and one of the top exchanges in Asia.

HISTORY

Scientists believe that communities of hunters and gatherers lived on the Japanese islands at least 6,500 years ago. By 200 B.C., farming had developed, and the people lived in villages.

From the A.D. 200's, Japan was controlled by warring clans. The Yamato clan established its authority in the 400's. Under the Yamato, Japan borrowed many ideas and technologies from China. Under the Taika Reform program in 646, Japan also adopted many features of China's centralized imperial government.

In 794, the Japanese capital was established at Heian, later called Kyoto. The Fujiwaras, a powerful noble family, gained control of the emperor and his court in 858 and ruled Japan for about 300 years. During this time, the emperors officially reigned, but they lost all real power.

Powerful bands of warriors called *samurai* fought for control of the imperial court during the 1000's. The Minamoto family headed one of these bands. In 1192, the emperor gave Minamoto Yoritomo the title *shogun* (general), and Yoritomo's military government in Kamakura became known as the *shogunate*. The shoguns controlled Japan until 1867, but always in the name of the emperor.

In 1543, Portuguese sailors became the first Europeans to reach Japan. Eventually, Roman Catholic missionaries from Portugal and Spain converted many Japanese to Christianity. However, Tokugawa Ieyasu, who established his shogunate in 1603, feared that the Christian missionaries might bring European armies with them to conquer Japan. As a result, the Tokugawa government cut ties with other nations during the 1630's, and Japan was isolated from the rest of the world for more than 200 years.

In 1853 and 1854, Commodore Matthew C. Perry of the United States sailed warships into the bay at Edo (Tokyo) and presented U.S. demands to Japan. Later, the Tokugawa government signed a treaty

Persecution of Christians and foreigners marked the beginning of Japan's isolation in the 1600's. The Tokugawa government believed that domestic order depended on preventing all contact with the outside world. Japanese were not allowed to leave the country, and some foreign sailors who had been shipwrecked on Japan's shores were killed.

granting the United States trading rights in two Japanese ports. Other European nations soon signed similar treaties with Japan.

In 1867, the Tokugawa shogunate was overthrown, and the emperor regained his traditional power. In 1868, Emperor Mutsuhito moved Japan's capital from Kyoto to Tokyo and announced the return of imperial rule. Emperor Mutsuhito adopted the title *Meiji*, meaning *enlightened rule*. During the Meiji period, from 1867 to 1912, Japan developed into a modern industrial and military power.

Japan soon expanded its territory as a result of three wars: the first Chinese-Japanese War (1894-1895), the Russo-Japanese War (1904-1905), and World War I (1914-1918). By 1918, the Japanese Empire controlled Taiwan, Korea, the southernmost tip of Manchuria, and several Pacific islands.

In 1931, Japan seized all of Manchuria, and open warfare between Japan and China began in 1937. Meanwhile, Japan had signed anti-Communist pacts with Nazi Germany and Fascist Italy. At the outbreak of World War II (1939-1945) in Europe, Japan allied itself with Germany and Italy. In 1941, Japan attacked U.S. bases at Pearl Harbor in

TIMELINE

4500 B.C.	Culture consists of hunters, gatherers.
660 B.C.	Per legend, Tenno becomes first emperor.
200's B.C.	Farming practiced, and people live in villages.
A.D. 200-400	Warring clans (related families) control Japan.
400's	Japanese adopt Chinese writing system and calendar.
646	Taika Reform sets up central government controlled by emperor.
858	Fujiwara family gains control of imperial court.
1192	Yoritomo becomes first shogun.
1543	Portuguese sailors become first Europeans to reach Japan.
1603	Tokugawa shogunate founded.
1630's	Japan begins isolation.
1853-1854	Commodore Perry of the United States opens two Japanese ports.
1867	Tokugawa shogunate overthrown. Emperor regains traditional powers.
1868	Mutsuhito announces Japan's intention to become a modern industrial nation.
1894-1895	Japan defeats China.
1904-1905	Japan is established as a world power following victory in Russo-Japanese War.
1914	Japan enters World War I on the Allied side.
1923	Earthquake destroys much of Tokyo and Yokohama.
1931	Japan seizes Manchuria.
1937	Japan goes to war against China.
1941	Japan attacks U.S. bases at Pearl Harbor.
1945	Japan surrenders in World War II after atomic bomb attacks on Hiroshima and Nagasaki.
1947	New constitution takes effect.
1950's	Akira Kurosawa becomes first Japanese motion-picture director to gain international fame.
1951	Japan signs security treaty with United States.
1952	Allied occupation of Japan ends.
1989	Emperor Hirohito dies. He is succeeded by Crown Prince Akihito.
1994	Tomiichi Murayama becomes Japan's first socialist prime minister since 1948.
2001	Junichiro Koizumi becomes prime minister.
2011	A powerful earthquake followed by a tsunami cause the deaths of at least 15,800 people.

Tokugawa Ieyasu (1542-1616)

Emperor Mutsuhito (1852-1912)

Junichiro Koizumi (1942-   )

Hawaii in order to bring the United States into the war as well.

The war came to an end in 1945 after the United States dropped atomic bombs on the Japanese cities of Hiroshima and Nagasaki in August. Japan suffered a crushing defeat in the war. Japanese casualties numbered in the millions, and much of the country had been destroyed. Japan lost all its territories, keeping only its four main islands and small islands nearby. Allied military forces occupied Japan until 1952, after a security treaty had been signed with the United States in 1951.

A new Japanese constitution took effect in 1947, providing the basis for the Japanese government of today.

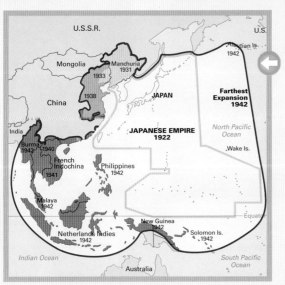

The expansion of the Japanese Empire—which had begun in the late 1800's—accelerated with conquests of the East Asian mainland in the 1930's. The empire continued to expand during World War II when Japan took much of Southeast Asia and many Pacific islands.

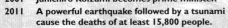

The wedding of Prince Aya, Emperor Akihito's son, to Kiko Kawashima took place in 1990. Like his father, Prince Aya broke with tradition and married a woman who did not belong to the Japanese nobility.

THE SAMURAI

By about A.D. 858, aristocratic families began establishing great private estates in the Japanese countryside. By the 1000's, these estates had become increasingly independent. The lords who controlled the estates were called *daimyo.* They hired bands of warriors to protect their lands and the peasants who worked them. These warriors became known as *samurai.*

The samurai developed from bands of warriors called *bushidan,* who were kinsmen of the estate lords they served. The bushidan consisted of loosely formed armies that usually dissolved after fighting a particular battle. After the rise of the feudal system, however, the bushidan became much more tightly organized. They were composed not only of members of the daimyo's family but also of men who were not related by blood. By the time of the Tokugawa shogunate in 1603, the term *samurai* referred to the entire military class, including the samurai warriors, the daimyos, and the shogun. About 5 percent of Japan's population belonged to this elite class.

The Bushido code

A code of unquestioning obedience and loyalty, called *Bushido,* bound the samurai warriors to their lords. According to this code, a samurai was expected to lay down his life for his daimyo. And, prizing his honor above his life, a samurai atoned for dishonor by committing *hara-kiri,* or ceremonial suicide.

In service to their daimyo, the samurai warriors wore two swords—the *katana,* a long sword with a slightly curved blade, and the *wakizashi,* a shorter sword. The samurai also carried a weapon called a *naginata,* a blade mounted on a long pole. Samurai warriors usually engaged in hand-to-hand combat using their swords or naginata.

In the late 1500's, Hideyoshi, a great warrior who controlled Japan at that time, ruled that only the samurai were allowed to carry and use weapons. Until this time, soldiering had been combined with farming and other professions, and even Buddhist temples contained their own arsenals. Hideyoshi carried out a series of raids to disarm farmers and Buddhist monks. This move made the military into a distinct class.

Craftsmen fashion the blade of a traditional samurai sword. The two swords that symbolized the samurai's status as a warrior were worn and used with much ritual.

The Jidai Matsuri, which means Festival of the Ages, is celebrated in Kyoto every October. At this event, people dressed in historical costumes—including that of the samurai—parade through the city.

Although some samurai resigned themselves to their loss of status and found new occupations, many others were discontented and became involved in uprisings. In 1877, these samurai launched the Satsuma Rebellion, a large-scale revolt against the government. The samurai suffered crushing defeat.

Nevertheless, remnants of the samurai tradition continued to endure until World War II, when more than 1,000 Japanese pilots flew *kamikaze* (divine wind) suicide missions against Allied warships. The kamikaze pilots, who crashed planes filled with explosives, considered it an honor to die for their emperor, believing that he ruled Japan by divine right. After the emperor surrendered in 1945, many Japanese committed hara-kiri in front of the Imperial Palace in Tokyo.

The samurai were graded in military ranks, each with an appropriate income paid in rice. A samurai retained his status as long as he remained in the service of his daimyo. But once his lord was overthrown, the samurai became a *ronin,* a warrior without a lord. By the 1600's, large numbers of these lordless warriors were working in nonmilitary occupations, and by 1700 most of the ronin had taken administrative posts and had lost both the will and the skill to fight effectively.

Modern samurai

The restoration of imperial rule in 1867 and the rapid modernization of Japan marked the end of the samurai. After the Japanese abolished feudalism in 1871, the samurai lost their privileges as a distinct class. Also, the government ruled in 1876 that the samurai could not wear their two traditional swords in public—a tremendous blow to the samurai's sense of honor.

Samurai warriors wore distinctive headdresses and magnificent protective armor that identified the family the warrior served.

KYOTO

Emperor Kammu established Japan's capital at Kyoto on Honshu Island in A.D. 794. He called it *Heiankyo*, meaning *capital of peace and tranquility*. Many Japanese called it *Miyako*, meaning *imperial city*, or *Kyoto*, meaning *capital city*.

The first 400 years of the capital's history—from 794 to 1185—was a time of great cultural achievement called the Heian period. It marked the end of centuries of cultural borrowing from China and the beginning of distinct Japanese styles in literature, art, religion, and other areas. These styles differed greatly from those of the Chinese.

During the Heian period, the greatest literature was written by women. Sei Shonagon, a lady in waiting to a noblewoman, portrayed Heian court life around the early 1000's in her volume of essays called *The Pillow Book*. Another book about court life, *The Tale of Genji* by Murasaki Shikibu, is the finest achievement of Heian literature. Indeed, many consider this lengthy novel to be the greatest work of Japanese fiction. Written around 1010, Murasaki's book surpassed previous stories with its sophisticated style and insightful descriptions of human emotions.

Another important development of the early Heian period was the introduction of Japanese characters called *kana*. Until this time, the Japanese language had been written in Chinese characters. The use of kana allowed a new freedom of expression for the Japanese, and some of the Heian period's finest literary works were written in this new script.

Although Buddhism declined in China during this period, it flourished in Japan. Shinto, Japan's traditional religion, was merged to a great ex-

A bamboo grove stands in Kyoto near the Tenryu-ji Buddhist temple. The name means "Temple of the Heavenly Dragon."

The dance of the crane, a symbol of long life, is part of the Gion Festival held at Yasaka Shrine in Kyoto. The festival, which dates from the 800's, was originally held to honor a Shinto god of good health.

Lovely gardens surround the teahouse of Kyoto's Imperial Palace, which was built in 794 and rebuilt in 1855.

A Buddhist temple in Kyoto, set amid trees and gardens, combines the Buddhist principle of quiet contemplation with the Shinto reverence for nature.

The Golden Pavilion, built by the shogun Yoshimitsu in 1397 and rebuilt in the 1950's, is one of the most beautiful sights in Kyoto. Gold leaf covers parts of the structure, which served as both a villa and a temple. The pavilion was reconstructed according to the original plan following a fire.

tent with Buddhism. Shintoists worship many gods, called *kami,* which they believe inhabit mountains, trees, and other elements of nature. During the Heian period, Shintoists identified Buddhist gods as kami and used Buddhist images to represent the kami in their shrines. Kyoto became an important religious center, with many Buddhist temples and Shinto shrines housing priceless works of art.

In 1185, however, the Minamoto family gained control of the imperial court at Kyoto—an event that marked the end of imperial culture and ushered in nearly 700 years of military domination under shogun rule. Tokyo replaced Kyoto as the nation's capital in 1868.

Today, Kyoto is one of Japan's largest cities. Although Kyoto was the only major Japanese city to escape bombing during World War II, earthquakes, fires, and floods have destroyed most of the buildings, temples, and shrines built during the Heian period. However, many of these ancient structures have been rebuilt in their original style, and the layout of the city still preserves Kyoto's original grid pattern, with its intersecting north-south and east-west avenues. As a result, many modern streets follow the routes of the ancient city of Kyoto.

TOKYO

During most of its history, Tokyo was called *Edo*. Historians believe that a powerful family first lived in the area around 1180. Its location overlooking the wide Kanto Plain and Tokyo Bay gave the area military importance, and in 1457 a powerful warrior named Ota Dokan built a castle there. Edo developed around the castle.

However, Edo's development into Japan's chief city did not begin until 1590, when Tokugawa Ieyasu made it his headquarters. Edo became Japan's political center in 1603, after Ieyasu became shogun of Japan. When shogun control ended in 1867, Emperor Mutsuhito transferred the capital from Kyoto to Edo and moved into the Edo castle. Edo then was renamed *Tokyo*, which means *eastern capital*.

After 1868, Japan—and especially Tokyo—rapidly adopted Western styles and inventions. European architects and building styles were employed to develop the city, and little attempt was made to preserve wooden buildings constructed in the ancient Japanese architectural style. By the late 1800's, Tokyo had begun to look in many ways like a Western city.

The city had to be rebuilt following a massive earthquake in 1923, which destroyed most of central Tokyo. About 59,000 people died in the disaster. World War II brought destruction to Tokyo once again. More than 250,000 people were killed or listed as missing, and about 97 square miles (251 square kilometers) of Tokyo were ruined. The people of Tokyo began to rebuild their city after the war, but without much planning. Buildings went up wherever there was room, and as a result, many of Tokyo's structures are modern. Most of the remaining buildings in Japanese style are religious shrines or temples.

Tokyo today is Japan's major business and cultural center, as well as the home of the Japanese

Pedestrians cross the street in Akihabara, a district in central Tokyo famous for its many electronics shops. The district is popularly known as Electrical Town. The shops sell computers, cameras, television sets, mobile phones, and home appliances as well as second-hand goods and electronic junk.

emperor and the headquarters of the national government. The city is also a major tourist center. The Imperial Palace, the home of the Japanese emperor, and the Meiji Shrine, which lies about 3 miles (5 kilometers) southwest of the palace, draw tourists from all over the world. Many Japanese, dressed in traditional garments, visit the well-known shrine on New Year's Day. Tokyo's parks, with their spring displays of cherry blossoms and Japanese-style gardens, are also popular with tourists. They not only reflect the Japanese love of beauty, but also offer relief from the crowds in a city that is now one of the most densely populated places on Earth.

With about 8-1/2 million people, Tokyo is one of the largest cities in the world. Tokyo, also called the *city proper*, is part of a huge urban area that also includes the cities of Yokohama, Chiba, and Kawasaki. This area is the largest urban center in the world, with an estimated population of more than 30 million people.

Japan's Shinkansen, or bullet trains, make up a large fleet of high-speed superexpress trains that run the length of the main island of Honshu.

The Imperial Palace lies at the heart of the city center. Northeast of the palace is the Kanda district, famous for its bookstores, and the Asakusa district, featuring many of the city's restaurants and theaters. Farther north lies popular Ueno Park, which offers a variety of cultural attractions, including a zoo and an art museum. Other beautiful parks are scattered throughout the city.

Tokyo's Imperial Palace consists of several low buildings set in beautiful parklike grounds. Stone walls and a series of moats separate the palace from the rest of the city.

Tokyo's soaring population has created a serious housing shortage, and the city government has begun financing the construction of low-rent housing projects.

However, most of Tokyo's housing developments are located far from the city proper, and workers who live in outlying areas spend up to four hours a day traveling to and from their jobs in downtown Tokyo. Severe traffic jams on the city's highways occur frequently. Millions of people cram aboard the city's commuter trains, which are so crowded that employees called *pushers* are hired to shove passengers into packed trains.

In spite of Tokyo's crowded conditions, the city has relatively little crime and poverty. Because of Tokyo's strong economy, most people can find jobs. In addition, the local and national governments provide aid for those who cannot support themselves.

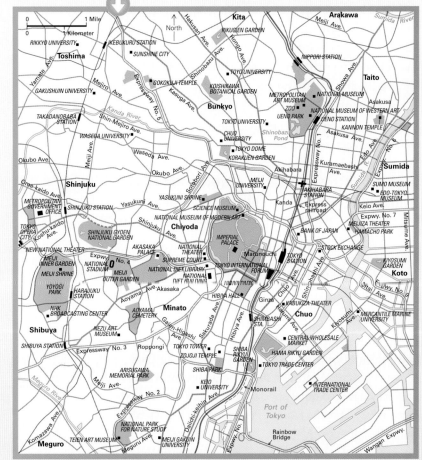

PEOPLE

The Japanese people of today are probably descended from the Yayoi, a people who established agricultural villages in what is now Japan around the 200's B.C. Chinese, Koreans, and a group of people called the Ainu make up the largest minority groups in Japan. Some scientists believe that the Ainu were Japan's original inhabitants. Today, most of these people live on Hokkaido, the country's northernmost island.

About two-thirds of the Japanese people live in urban areas. Much of the population is concentrated in three major metropolitan areas: the Tokyo metropolitan region, which also includes the cities of Chiba, Kawasaki, and Yokohama; Osaka; and Nagoya—all on the island of Honshu. Most Japanese city dwellers are employed in factories, businesses, the government, and service industries and enjoy a comfortable standard of living.

Only about one-third of the Japanese people live in rural areas. Most of the rural population work as farmers, but some Japanese make their living by fishing and harvesting edible seaweed along the coasts. Although their standard of living has increased steadily since World War II, most rural workers do not earn as much as city dwellers. Since the late 1950's, many rural Japanese, particularly young people, have moved to urban areas to seek better-paying jobs.

East meets West

Life in Japan's big cities combines the old and the new, as well as the East and the West. High-rise office buildings dominate the commercial districts, and modern transportation systems carry millions of commuters each day. But many old customs still flourish. For example, numerous big-city shops specialize in traditional items, such as straw mats called *tatami,* which are used as floor coverings. In addition, even the most crowded cities have beautiful gardens, parks, and shrines—all of which reflect the traditional Japanese love of nature.

Although most of the country's city dwellers wear Western-style clothing in public, some Japanese people—particularly the older generation—still dress in the traditional *kimono* at home. Almost all Japanese wear kimonos for festivals, holidays, and other special occasions. Worn by both men and women, the kimono is tied around the waist with a sash called an *obi.*

Traditional garments are part of the wedding ceremony for many Japanese couples.

Children play in a snowhouse during the celebration of the Snow Festival in Sapporo, Hokkaido. Although Japanese parents are traditionally strict, a number of festivals are held especially for children.

Commuters crowd into a train on the Yamanote Line, Tokyo's most important train line. The circular line connects Tokyo's major city centers.

City housing, also a blend of East and West, includes both modern apartment buildings and traditional Japanese houses. Most traditional houses feature lovely gardens, graceful tile roofs, and sliding paper screens between rooms. Tatami cover the floors, and people sit on cushions and sleep on padded quilts called *futons*. However, many Japanese apartments and houses have one or more rooms with Western-style furniture and carpets on the floors.

Family life

Before 1945, many Japanese lived together in large extended families and followed strict social customs. Husbands had complete authority over their wives, and children were expected to obey their parents without question. Parents even selected their children's marriage partners—a bride and groom often met for the first time on their wedding day.

Today, most Japanese live in smaller family units consisting of only parents and children, and relationships within these families have become more democratic. Children are given much more freedom, and most young people select their own marriage partners. Women, too, guaranteed equal rights by the Constitution of 1947, are no longer dominated by their husbands or male relatives. As a result, an increasing number of Japanese women now work outside the home.

Nevertheless, although the urbanization of Japan has produced many changes in the country, the Japanese people still maintain strong family ties and a deep respect for authority.

A businessman proudly displays pearls strung in his factory. The pearls come from oysters cultivated and harvested in Japan's coastal waters.

A New Year's parade goes on in spite of a snowstorm. Many Japanese dress for the occasion in their most elaborate and colorful kimonos.

Seafood and rice constitute the traditional diet of the Japanese people. But meat and dairy products have become more popular since the 1950's, and many Japanese substitute bread for rice at some meals.

ETHNIC AND SOCIAL GROUPS

Chinese, Koreans, and Brazilians make up the largest ethnic minority groups in Japan. Groups of Chinese and Koreans are scattered throughout Japan, though large Korean communities can be found in Osaka, Kobe, and Kyoto. Koreans in Japan who have declared their loyalty to North Korea belong to an organization called Chosen Soren. This organization, which is supported by the North Korean government, runs its own schools, where only the Korean language is spoken.

The Ainu people are another noteworthy minority group in Japan. Scientists are uncertain about the origin of the Ainu, who may have been Japan's original inhabitants. Some anthropologists think the Ainu are related to European peoples, while others believe they are related to Asian peoples or to the original inhabitants of Australia. Over the centuries, many Ainu have intermarried with the Japanese and have adopted Japanese customs. Today, only a small number of Ainu follow their traditional way of life. These Ainu live in isolated communities on Hokkaido Island.

Government aid

The Ainu have long been victims of discrimination. However, they have carried on a movement to achieve fair treatment, and the government has instituted a program to aid them economically. As a result of these measures, prejudice against the Ainu people has decreased.

The Korean minority in Japan also has faced discrimination. In 1990, however, government leaders from both South Korea and North Korea demanded equal treatment for the Koreans in Japan. As a result, Japan agreed to grant Koreans more civil rights.

The ramshackle houses of the burakumin are usually clustered in the poorer sections of Japan's urban areas. The buildings' inferior construction offers little resistance to the country's frequent earthquakes.

Shoe shining and other low-paying jobs, such as tanning leather and working in prisons and slaughterhouses, are often the only ones open to the burakumin.

Outcasts and criminals

The people who have suffered the greatest injustice in Japan are a group of Japanese known as the *burakumin* or *eta*. In Japanese feudal society, the burakumin were outcasts because they came from villages associated with tasks considered unclean by Buddhist doctrine. These tasks included the execution of criminals, the slaughter of cattle, and the tanning of leather. According to Buddhism, not only the tasks, but also the people who performed these duties, were unclean.

Feudalism was abolished in Japan in 1871, but the outcasts remain. Today, the burakumin make up about 2 percent of Japan's population. Though they are not ethnically different from other Japanese, they still suffer from discrimination. Many live in segregated urban slums or special villages. The burakumin have started an active social movement to gain fair treatment but have achieved only limited success.

The *yakuza,* who also live outside the mainstream of Japanese society, are members of a criminal organization. They form a significant minority in Tokyo and other large cities. Because yakuza members band together in families or clans, they are sometimes called the "Japanese Mafia." Some yakuza families claim kinship with the ronin, the lordless samurai of the 1600's.

With their colorful tattoos, the flamboyant yakuza have provided a popular subject for comics and movies about underworld gangsters. In the past, the yakuza were thought to be relatively harmless to society. However, their involvement with drug dealing has led the Japanese to regard them as a threat.

A vagrant stops for a rest outside a Tokyo shop. Although most people in Japan can find jobs, those without the proper ancestral background may encounter discrimination and have difficulty finding work.

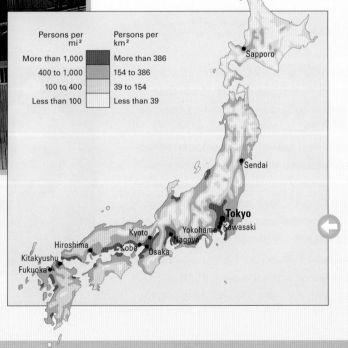

Persons per mi²	Persons per km²
More than 1,000	More than 386
400 to 1,000	154 to 386
100 to 400	39 to 154
Less than 100	Less than 39

Sapporo

Sendai

Tokyo

Kawasaki

Yokohama

Kyoto

Nagoya

Hiroshima

Kobe

Osaka

Kitakyushu

Fukuoka

Japan's population is unevenly distributed. About 90 percent of the people live on the coastal plains. The Pacific coast from Tokyo to Kobe is the most densely populated area. The Ainu have been pushed northward to Hokkaido, Japan's northernmost island.

POPULAR CULTURE

Popular culture in Japan is a blend of Japanese traditions and modern Western influences. Since the Meiji period, Japan has borrowed sports and arts from the Western world, particularly the United States. However, much of this imported culture has been adapted to suit Japanese tastes. For example, baseball, which was introduced into Japan in the late 1800's, is now one of the country's most popular spectator sports. Yet to the Japanese, the game represents an art form in which play is carried out with order and harmony. The slowness of the game allows the spectators to pay attention to its every detail, and the duel between the pitcher and batter provides drama.

While baseball successfully combines Japanese spirit with Western know-how, sumo wrestling is uniquely Japanese. The country's national sport, sumo is immensely popular in Japan. The wrestlers perform on a raised platform, which accentuates their huge size, and many rituals surround the tournaments.

Among the participant sports in Japan, rubber-ball baseball, golf, skiing, and tennis are popular. The traditional Asian martial arts of *aikido, judo,* and *karate* involve fighting without weapons. Another favorite pastime for Japanese of all ages is *pachinko,* a noisy type of pinball. A pachinko arcade contains hundreds of machines lined up in rows. There are thousands of such arcades throughout Japan.

Motion pictures

Early Japanese films were heavily influenced by several forms of traditional Japanese drama. Silent films of the early 1900's employed a narrator who kept the audience informed of the plot—a custom derived from the Japanese *puppet* theater, which developed during the late 1600's. Some films borrowed the choruses and musical accompaniment of the *noh* play, developed during the 1300's. Films featuring the stylized enactment of historical and domestic events were grounded in traditional *Kabuki* theater, which began in the 1600's.

Baseball is a year-round obsession with the Japanese. Many enthusiasts have compared the concentration required to play the sport to the concentration of Zen Buddhists during meditation.

Samurai films, one of the most popular styles of film in Japan, evolved from the Kabuki drama. Samurai films range from the artistic efforts of Akira Kurosawa, possibly the country's best-known film director, to action-packed swashbucklers that often combine swordsmanship with magic. However, the most popular films in Japan are movies about gangsters and professional gamblers—the so-called yakuza films. These films usually pit a "good" yakuza, who lives according to an established code of honor, against a gang of "bad" yakuza, who are willing to violate that code.

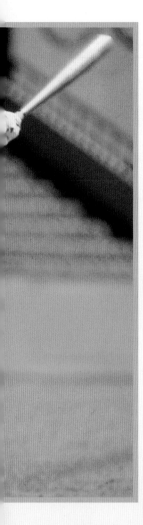

Popular music

Western styles have also influenced Japanese music. Popular songs in Japan, called *kayokyoku,* have borrowed heavily from many types of popular Western music, including pop, folk, rock, and heavy metal. However, the Japanese have developed their own brand of popular music sung by teenage singers, usually female, called *kawaiko-chan.* The kawaiko-chan accompany their songs with hand gestures and body movements especially choreographed for them. These singers are discovered, packaged, and promoted by production companies.

A major influence on Japanese popular music was the technological innovation called *karaoke* (empty orchestra). In karaoke, a vocalist sings into a microphone while a prerecorded tape plays background instrumental music. Karaoke is used in bars, and video versions are popular in many Japanese homes.

Perfecting their putting, Japanese businessmen exhibit their kinship with their Western counterparts. These players hone their skills on a practice area constructed on the roof of a Tokyo department store.

A massive sumo wrestler enters the ring, while a sumo official, dressed in traditional costume, stands at attention. Sumo tournaments attract large, enthusiastic crowds. Many matches are televised.

The influence of Western fashions is apparent among many young Japanese. Some even form groups that are identified by their tastes in clothing and music. Young men in one such group wear 1950's American clothing and pose and dance in public parks.

Pachinko players manipulate the machine's controls to direct a metal ball through a horizontal maze of obstacles. Winners are rewarded with extra balls that can be exchanged for prizes, or used to play again.

SHINTO AND BUDDHISM

Japan's oldest religion is Shinto—a native religion of Japan that dates from prehistoric times. *Shinto* means *the way of the gods,* and Shintoists worship many gods, or *kami,* which they believe are found in nature. The sun, water, mountains, trees, and stones have particular importance in Shinto worship. According to Shinto mythology, Japan's emperors are direct descendants of the sun goddess—the supreme kami. The other four elements of nature—water, mountains, trees, and stones—are prominent features throughout the Japanese landscape.

Shinto emphasizes rituals and ceremonies during which offerings are made to the kami, accompanied by prayers. On special occasions, festivals called *matsuri* are held to honor a particular god. Worshipers take part in solemn religious ceremonies during a matsuri, but they also entertain the god with food, drink, music, and dance, hoping to earn the god's good will. They believe that a god that is pleased by the festival will give them such gifts as long life, wealth, peace, good health, and a rich harvest.

These ceremonies and festivals generally take place at shrines, which range from a large complex of several buildings and gardens to a small space in someone's home. Roadside shrines dedicated to certain kami are also common. In addition, the kami themselves can be used as places of worship. Mountains, for example, have been worshiped as gods and used as religious shrines.

Shinto rituals and festivals celebrate the important occasions of life and the agricultural year, such as birth, marriage, and harvesttime. Shinto expresses a simple joy in life, but it does not deal with death. Buddhism gave the Japanese new insight into this aspect of the life cycle.

Buddhism was introduced to Japan from China and Korea about A.D. 552, when Japan was heavily under the influence of Chinese ideas. Generally, Buddhism teaches that people can achieve peace and be free from suffering only if they rid themselves of attachment to worldly things. The Japanese accepted the religion, but they added some of their own appreciation of life and nature to the Buddhist teachings. They also merged Buddhist art and culture with Shinto beliefs. Temples and monasteries were built to honor Buddha, but Buddhist images were also used to represent Shinto gods.

Priests celebrate Aoi Matsuri, a major Shinto festival, with music. Participants in the festival, dressed like nobles of the late Heian period, lead a procession to two Shinto shrines, where ceremonies are held.

A Buddhist monk seeks donations on a busy street in Tokyo. Modernization has not resulted in the abandonment of religion in Japan. On the contrary, because Buddhism and Shinto reflect the Japanese ideal of a harmonious society working together, religion has helped to unify and strengthen the country.

The ornate Toshogu Shrine at Nikko, built during the 1600's and dedicated to Shogun Tokugawa Ieyasu, contrasts sharply with the Japanese ideal of simplicity and harmony. During this period, the Japanese celebrated military power and material splendor.

The Great Buddha of Kamakura, serene and peaceful, sits cross-legged in meditation. Zen Buddhists also assume this position while meditating.

A wooden gate called a torii is the symbol of Shinto. A torii consists of two posts connected by crossbars. It stands at the entrance of a Shinto shrine. The posts represent pillars that support the sky, and the crossbars symbolize the earth. This torii, on a small island near Hiroshima, is partly submerged by the high tide.

A form of Buddhism called Zen, meaning *meditation* was introduced into Japan in the 1100's. Zen Buddhists believe meditation is the key to achieving a state of spiritual enlightenment called *satori*. The development of the traditional Japanese tea ceremony was highly influenced by both Zen and Shinto. Zen turned the making and drinking of tea into a time of intense meditation, while Shintoists found beauty and significance in its simplicity.

During the 1800's, many Shintoists began to reject Buddhism. In the mid-1800's, a movement to *State Shinto* stressed patriotism and the divine origins of the Japanese emperor. Japan's defeat in World War II shattered this movement. As a result, the government abolished State Shinto, and the emperor denied that he was divine.

Today, almost all of Japan's population practice a combination of Shinto and Buddhism. For example, they may celebrate births and marriages with Shinto ceremonies, but funerals are observed with Buddhist ceremonies.

JAPANESE GARDENS

The Japanese have been creating artfully land-scaped gardens since before the A.D. 500's. These gardens range from spacious parks created for the nobility to tiny tea gardens. Large or small, the gardens mirror the natural beauty of Japan's landscape and reflect the Japanese ideals of simplicity, harmony, and tranquillity.

A Japanese garden is designed to appear as if it were created by nature and not by a gardener's artistry. Japanese gardens are often miniature replicas of scenes found in nature or described in Japanese literature. Thus, familiarity with a specific literary work, such as *The Tale of Genji,* might be required to fully appreciate the perfection of a particular Japanese garden.

Some Japanese gardens are quite elaborate, with stone paths winding past trees, buildings, hills, and lakes. These elements are cleverly arranged to present a new and pleasing scene at every turn in the path. Gardens in more confined spaces may give the illusion of spaciousness. The landscape gardener often uses perspective to achieve this end. For example, large and eye-catching objects may be placed in the foreground, while rocks and plants of decreasing size fill the background. The gardener may also conceal or camouflage the boundaries of the garden to disguise its actual size. Thus, as paths and streams disappear behind rocks or trees, they appear to be winding off into an undefined distance.

Japanese gardeners also use a technique known as "borrowed scenery," in which a rock or shrub is taken from its natural surroundings and placed in a garden setting. A large stone removed from a mountain, for example, may be set in a garden to create the feeling that the mountain is part of the scenery. Some gardens achieve a similar effect by making the surrounding scenery a part of the garden. In the Fukiage Garden in Tokyo's Imperial Palace, for instance, the path winds past a terrace from which the visitor can view Mount Fuji.

Japanese gardens come in many forms, including the water garden, the dry garden, and the tea garden. The popularity of water gardens reflects the importance of water to the Japanese. In a water garden, lakes or ponds dotted with tiny islets surround graceful buildings. Nestled in a forest of trees, Kyoto's Golden Pavilion actually sits on a lake. The reflection of the pavilion in the water seems to draw nature into the building and draw the building into nature.

Among the many different kinds of Japanese gardens, the dry garden is perhaps the most unusual to Western eyes. Unlike the Western garden, with its colorful flower beds and grassy lawns, a dry garden contains only sand and rocks. The rocks, carefully chosen for their shape, color, and size, are arranged on sand that has been raked in geometric patterns.

The stone garden at Ryoanji Temple in Kyoto contains only rocks and sand. The sand is raked to represent the waves of the sea, and the rocks symbolize islands emerging from the water.

Trees surrounding a lake on the grounds of the Imperial Palace in Kyoto are the result of nature and the gardener's artistry. The trees have been carefully trimmed to create the most pleasing composition of shapes, color, and foliage.

Many dry gardens re-create a landscape in miniature. For example, the swirling patterns of sand in the dry garden at Daisenin Temple in Kyoto are designed to suggest a waterfall and an ocean, while the rocks set against the sand symbolize boats and bridges. By contrast, the famous rock garden at Ryoanji Temple in Kyoto is abstract in design, with 15 rocks set like islands on waves of white, raked sand. The dry garden creates a simple world of stillness and beauty that reflects the Buddhist ideal of serenity. It also draws on the Shinto belief in the spiritual forces of nature.

The tea garden, also inspired by Buddhism, is built at the approach to a Japanese teahouse, where a ceremonial cup of tea inspires quiet repose and refreshes the spirit. A traditional tea garden is very simple so that it does not distract the mind from the tea ceremony. Many tea gardens consist only of a short path of stepping stones leading to the teahouse. The stones, chosen for their size, shape, and texture, are placed close together along the path so that a person must concentrate on walking across the stones and is, therefore, forced to disregard the surroundings and concentrate on the ceremony.

Celebrating the appearance of the cherry blossoms and the coming of spring, Japanese women, dressed in traditional costumes with sashes the color of cherry blossoms, parade through a Tokyo street. Flower-viewing parties are held throughout Japan during this period, and many people picnic beneath the lovely cherry blossoms.

The gardens in Kyoto's Shugakuin Palace reflect the harmonious relationship between gardens and architectural structures. Screen doors on every side of the building open onto the garden, so that the garden appears to enter the building.

Jordan is an Arab kingdom in the heart of the Middle East. Much of Jordan's modern history has been shaped by events in the land to its west, the area once called Palestine. In fact, during the early 1900's, Jordan was called *Transjordan* because of its location across the River Jordan from Palestine.

The history of the region that is now Jordan goes back thousands of years. About 2000 B.C., Semitic nomads moved into the region, and by about 1200 B.C., four Semitic peoples—Ammonites, Amorites, Edomites, and Moabites—made a living there as farmers and traders. During the 900's B.C., the Israelite kings David and Solomon conquered the region, but Moabites led by King Mesha regained control about 50 years later.

A series of foreign invaders and rulers followed. In the 400's B.C., the area was controlled by the Nabataeans—traders who carved a capital city out of the rose-colored cliffs at Petra. In the 60's B.C., the Romans conquered the region. When the Roman Empire split in the late A.D. 300's, Jordan became part of the Byzantine Empire.

Muslims from the Arabian Peninsula defeated Byzantine armies in 636. The conquering Arabs brought the religion of Islam and the Arabic language to the people of the area. The Arabs established an important pilgrimage route through Jordan to Mecca, their holy city in Arabia.

About 1100, Christian crusaders from Europe captured land in the Middle East, including parts of Jordan, but the Muslim leader Saladin drove out the crusaders in 1187. Today, Saladin's shield, helmet, and eagle are displayed on Jordan's coat of arms.

Jordan was part of the Ottoman Empire during World War I (1914-1918), when Sherif Hussein of Arabia led an Arab revolt against Ottoman rule. With the help of the United Kingdom, the Arabs defeated the Ottomans in the Middle East.

The United Kingdom was then given the right to administer lands east and west of the River Jordan. In 1921, the land east of the river was named Transjordan and given partial self-government, with Hussein's son, Abdullah, ruling as *emir* (prince). The nation became independent in 1946 and was renamed Jordan.

Jordan became involved in the affairs of Palestine—the land west of the River Jordan—in 1948. In that year, part of Palestine became the new state of Israel, which was created as a homeland for Jewish people.

JORDAN

As soon as the new state was proclaimed, Jordan and other Arab countries attacked Israel. When the fighting ended in 1949, Israel occupied much of Palestine. Jordan took over the West Bank of the River Jordan and the eastern section of Jerusalem. Jordan's population more than tripled with the addition of about 400,000 Palestinian Arabs who lived on the West Bank and about 450,000 Palestinian refugees who moved from Israel into Jordan. This large Palestinian population caused political and economic problems. The Palestinians required food and shelter, and they competed with the Jordanians for power. In 1951, Palestinians assassinated Abdullah. In 1953, Abdullah's grandson, Hussein, took control as Jordan's king.

Increasing Arab-Israeli tensions led to the formation of the Palestine Liberation Organization (PLO) in 1964. Israel and the PLO staged raids against each other from time to time. Finally, on June 5, 1967, Israel attacked Jordan's ally Egypt. Jordan responded by attacking Israel. During the Six-Day War that followed, Jordan lost control of eastern Jerusalem and the West Bank. In addition to the influx of about 300,000 Palestinians from the West Bank, Jordan suffered from the loss of farmland and tourist income. Many Palestinians formed guerrilla groups to fight Israel. By early 1970, these forces had become almost a second government in Jordan and threatened to overthrow the king.

On Sept. 17, 1970, the Jordanian Army attacked the Palestinian guerrillas and defeated them within a month. But isolated fighting continued. At a meeting of Arab leaders in 1974, King Hussein agreed that the West Bank should become an independent Palestinian state if Israel withdrew from that region. Nevertheless, Jordan continued to help fund public services in the West Bank until 1988. In July 1994, Jordan and Israel signed a declaration that formally ended war between the two countries.

Hussein died in 1999. He was succeeded by his son Abdullah. In the 1990's, Israel withdrew from parts of the West Bank. But in 2002, it reoccupied most of those areas.

In 2011, antigovernment protests erupted in several Jordanian cities. The protesters called for the reduction of food and fuel prices and the removal of the prime minister. King Abdullah replaced the entire Cabinet.

283

JORDAN TODAY

Jordan is a land of many contrasts, where modern cities rise near ancient ruins. And, like its landscape, its government is a mix of past and present. Since 1946, the country has been an independent constitutional monarchy.

Government

Jordan is ruled by a king with wide powers. The king appoints a prime minister to head the government. He also appoints the members of the Council of Ministers, Jordan's cabinet, and the members of the Senate. The Senate is one house of the National Assembly, Jordan's legislature. Senators serve four-year terms. The second house of the legislature is the Chamber of Deputies. Its members are elected by the people to four-year terms. In addition, the king names a governor to head each district in Jordan and appoints all the judges in the country.

The government owns and operates the nation's radio and television stations and closely controls all communications. Nevertheless, there is more freedom of speech in Jordan than in many other Middle Eastern countries.

Economy

Jordan has a developing economy based on free enterprise. The nation depends on foreign aid and on the economy of the Middle East region as a whole. Many Jordanians work outside the country in neighboring oil-producing Arab nations and send money home to their families.

Within Jordan, service industries make up the largest part of the economy. They employ about 70 percent of the country's workers and account for more than 60 percent of the total value of the nation's economic production. Many service workers are employed by the government, while others work in banking, education, insurance, trade, tourism, or transportation.

FACTS

Official name:	Al Mamlakah al Urdiniyah al Hashimiyah (Hashemite Kingdom of Jordan)
Capital:	Amman
Terrain:	Mostly desert plateau in east, highland area in west; Great Rift Valley separates east and west banks of the Jordan River
Area:	34,495 mi² (89,342 km²)
Climate:	Mostly arid desert; rainy season in west (November to April)
Main river:	Jordan
Highest elevation:	Jabal Ramm, 5,755 ft (1,754 m)
Lowest elevation:	Shore of the Dead Sea, about 1,381 ft (421 m) below sea level
Form of government:	Constitutional monarchy
Head of state:	Monarch
Head of government:	Prime minister
Administrative areas:	12 muhafazat (governorates)
Legislature:	Majlis al-'Umma (National Assembly) consisting of the Senate with 40 members serving four-year terms and the Chamber of Deputies with 110 members serving four-year terms
Court system:	Court of Cassation (Supreme Court)
Armed forces:	100,500 troops
National holiday:	Independence Day - May 25 (1946)
Estimated 2010 population:	6,361,000
Population density:	184 persons per mi² (71 per km²)
Population distribution:	83% urban, 17% rural
Life expectancy in years:	Male, 74; female, 77
Doctors per 1,000 people:	2.4
Birth rate per 1,000:	28
Death rate per 1,000:	4
Infant mortality:	21 deaths per 1,000 live births
Age structure:	0-14: 36%; 15-64: 61%; 65 and over: 3%
Internet users per 100 people:	24
Internet code:	.jo
Languages spoken:	Arabic (official), English
Religions:	Sunni Muslim 92%, Christian 6%, other 2%
Currency:	Jordanian dinar
Gross domestic product (GDP) in 2008:	$20.01 billion U.S.
Real annual growth rate (2008):	5.8%
GDP per capita (2008):	$3,441 U.S.
Goods exported:	Clothing, fertilizer, gold, pharmaceuticals, phosphates, vegetables
Goods imported:	Crude oil and petroleum products, food, iron, machinery, motor vehicles
Trading partners:	China, Germany, India, Iraq, Saudi Arabia, United States

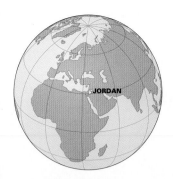

Jordan, an Arab kingdom in the Middle East, borders Syria, Iraq, Saudi Arabia, Israel, and the West Bank. Amman is Jordan's capital and largest city.

Many Jordanian people work for the military. In the face of wars with Israel in the 1960's and 1970's, as well as threats from Palestinians within the country, Jordan's government maintains large armed forces. But the cost of doing so is high and places a burden on Jordan's economy.

Jordan has few natural resources. Workers mine phosphates and potash, but the country lacks the valuable oil found in neighboring Arab lands. Jordan has only one dam for producing its own electricity and so must import oil to meet its energy needs. And because Jordan is mostly desert, only about 3 percent of the land is farmed.

Together, manufacturing and mining employ about 20 percent of Jordan's workers and account for about 30 percent of the nation's total economic production. Large plants refine petroleum and produce fertilizer and cement. Smaller plants produce batteries, ceramics, cigarettes, detergents, food products, pharmaceutical products, shoes, and textiles.

With the help of modern farming methods, many farmers along the River Jordan grow citrus fruits and vegetables. East of the river valley, farmers cultivate such grains as barley and wheat and such fruits as grapes and olives, as well as vegetables and nuts.

Jordan has a good transportation system with paved highways that link the country to its neighbors. The country is landlocked except for a tiny stretch of land in the southwest, on the Gulf of Aqaba. There, Al Aqabah, Jordan's only port, has been well developed to handle cargo.

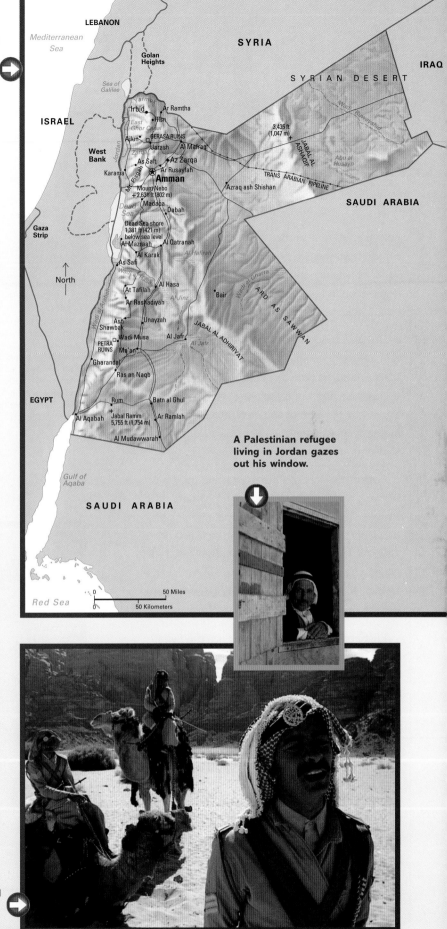

A Palestinian refugee living in Jordan gazes out his window.

Jordanian police on camelback patrol the desert town of Rum. Like many Middle Eastern countries, much of Jordan's land is desert.

LAND AND PEOPLE

Jordan has a varied landscape of deserts, mountains, deep valleys, and rolling plains. The nation's land area is divided into three main regions: the Jordan River Valley, the Transjordan Plateau, and the Syrian Desert.

The Jordan River Valley—a deep, narrow valley near the western border—extends along the River Jordan from the Sea of Galilee to the Dead Sea. The valley is actually part of the Great Rift Valley, a deep depression in the Earth's surface that extends well into Africa.

Because the Jordan River Valley receives very little rain, it was generally unsuitable for farming until the 1960's. At that time, the country developed an irrigation system and began to use plastic-covered hothouses to grow fruits and vegetables. These modern methods have made the river valley Jordan's major agricultural area.

East of the Jordan River Valley and the Dead Sea, the land rises steeply to form the Transjordan Plateau. This wedge-shaped plateau begins at Jordan's northern border, narrowing as it extends southward to the region around Maan. The high, rolling plains of the Transjordan Plateau are cut by steep *wadis,* or dry valleys. This area includes Jordan's largest cities as well as most of its farmland.

East and south of the plateau lies the Syrian Desert, Jordan's third major region. Hot and dry, the Syrian Desert is the northern part of the vast desert area that covers much of the Arabian Peninsula.

Most of Jordan's nearly 6.4 million people live in the northwest on the fertile high plateau. About 50 percent of the people are native Jordanians. Most of the others are Palestinian Arabs.

Most of these Palestinians fled to Jordan as refugees after the Arab-Israeli wars of 1948 and 1967. Others moved from the West Bank to the capital city of Amman between these wars. About 10 percent of Jordan's people live in crowded Palestinian refugee camps set up by the United Nations (UN). The UN also operates schools for the refugees.

A Bedouin entertains his visitors with music and tea at an encampment in the desert. Since the 1950's, many of these nomadic people have settled in rural villages or towns. Today, Jordan is largely an urban society.

The rolling plains of the Transjordan Plateau provide most of the country's farmland, though the Jordan River Valley is now a more productive agricultural area.

An ancient Roman amphitheater lies in the center of Amman, Jordan's capital and largest city. Romans took control of Jordan in the 60's B.C. and built a vast trading center at what is now Amman.

The ancient city of Gerasa, at present-day Jarash, near Amman, became an important Roman trading center after the A.D. 60's. Towering columns and stone avenues remind visitors of Jordan's 300 years as a Roman colony.

The great majority of Jordanians—more than 90 percent—are Muslims. Almost all follow the Sunni, or orthodox, branch. Islam affects every aspect of life for Jordanian Muslims. Devout Muslims pray five times a day, attend a mosque, fast, give to the poor, and, at least once in their lives, make a pilgrimage to the sacred city of Mecca in Saudi Arabia.

Though Arabic is the official language of Jordan, English is widely taught and spoken, and the government prints many documents in both languages. Ethnic minorities often speak their own language.

About four-fifths of Jordan's people live in towns or cities. Almost all the houses and apartments have electricity and running water, but some Jordanians have to live in dense, crowded neighborhoods. Still, living conditions in Jordan are generally better than in many other developing countries.

Most rural Jordanians live in village houses made of stone and mud or concrete. Many villagers grow crops and raise goats and chickens. Others work in construction or mining.

Less than 2 percent of Jordan's people are Bedouins—Arab nomads who live in tents and roam the desert with their camels and sheep in search of water and pasture. Since the mid-1900's, many of these nomadic people have settled in towns and villages.

Most Jordanians wear modern Western clothing, but men often cover their heads with a cloth called a *kaffiyeh*. Many women wear long, loose-fitting dresses. Some rural Jordanians, including the Bedouins, wear traditional robes.

Jordanians enjoy getting together at large family gatherings and picnics. They eat a variety of foods, including cheese, cracked wheat, flat bread, rice, vegetables, and yogurt. Chicken is popular, as is lamb, which is cooked in yogurt and served on a large tray of rice in a traditional Jordanian dish called *mansef*.